Spaces of Surveillance

Susan Flynn · Antonia Mackay
Editors

Spaces of Surveillance

States and Selves

palgrave
macmillan

Editors
Susan Flynn
School of Media
University of the Arts London
London, UK

Antonia Mackay
Department of English and Modern
  Languages
Oxford Brookes University
Oxford, UK

ISBN 978-3-319-49084-7   ISBN 978-3-319-49085-4   (eBook)
DOI 10.1007/978-3-319-49085-4

Library of Congress Control Number: 2017937509

Cover illustration: maja/Alamy Stock Photo

Printed on acid-free paper

This Palgrave Macmillan imprint is published by Springer Nature
The registered company is Springer International Publishing AG
The registered company address is: Gewerbestrasse 11, 6330 Cham, Switzerland

CONTENTS

v

Editors and Contributors

About the Editors

Susan Flynn is a lecturer at the University of the Arts, London where she specialises in contemporary media culture, digital and body theory and media equality. Her work is featured in a number of international collections and journals such as *American, British and Canadian Studies Journal* and *Ethos: A Digital Review of the Arts, Humanities and Public Ethics.*

Antonia Mackay is an Associate Lecturer at Oxford Brookes University and Visiting Lecturer at Goldsmiths, University of London. She has taught on a wide range of undergraduate and postgraduate modules including American Theatre, American Vistas, Critical Theory, Narrative and Narratology and Special Options in experimental Avant Garde and Twentieth Century Writing. She has published articles on Manhattan Maleness and Cold War ideology, as well as articles on space, technology and identity and won the Nigel Messenger Teaching Award at Oxford Brookes in 2014 and 2016.

Contributors

Simon Bacon Poznan, Poland

Tapo Chimbganda York University in Toronto, Toronto, Canada

Amy Christmas Qatar University, Doha, Qatar

Jeffrey Clapp Department of Literature and Cultural Studies, Education University of Hong Kong, Hong Kong, China

Francesca D'Amico York University in Toronto, Toronto, Canada

Susan Flynn University of Arts, London, UK; School of Media, University of the Arts London, London, UK

Alison Lutton University of Oxford, Oxford, UK

Antonia Mackay Department of English and Modern Languages, Oxford Brookes University, Oxford, UK

William Thomas McBride Illinois State University, Normal, USA

Jaclyn Meloche Art Gallery of Windsor, Windsor, Canada

Caleb Andrew Milligan University of Florida, Gainesville, USA

Frances Pheasant-Kelly Wolverhampton University, Wolverhampton, UK

Virginia Pignagnoli University of Turin, Turin, Italy

Mary Ryan Department of Political Science, Virginia Tech, Blacksburg, USA

Sam Tecle York University in Toronto, Toronto, Canada

Yafet Tewelde York University in Toronto, Toronto, Canada

List of Figures

CHAPTER 1

Introduction

Susan Flynn and Antonia Mackay

In 1948, George Orwell's *Nineteen Eighty-Four* portrayed a bleak future for mankind. A future almost entirely filled with surveillance technologies and surveilling practices; one where all seeing eyes and ears threaten to destroy individualism in favour of blind conformity. Orwell's ominous vision of Big Brother and the control exerted by the Ministries of Oceania, is one premised on the notion of surveillance, and with it, states of selfhood where the system "controls matter because we control the mind" (Orwell 1948, p. 268). The grim reality of Orwell's Big Brother is not merely the weight of sheer political power, but rather, the effects of surveillance on those who are watched, and thereby, those who are policed. By extension, as the novel's protagonist Winston Smith informs us, the power of these technologies to maintain the societal system around him, indeed Oceania itself, means this is not only a surveilled space, but a space of surveillance where "BIG BROTHER IS WATCHING YOU" (Orwell 1948, p. 3). The importance of surveillance technology in *Nineteen Eighty-Four* cannot be

S. Flynn
School of Media, University of the Arts London, London, UK
e-mail: susan.flynn.1@ucdconnect.ie

A. Mackay (✉)
Department of English and Modern Languages, Oxford Brookes University, Oxford, UK
e-mail: antoniamackay@brookes.ac.uk

© The Author(s) 2017   1
S. Flynn and A. Mackay (eds.), *Spaces of Surveillance*,
DOI 10.1007/978-3-319-49085-4_1

underestimated, for here is a system that can both watch and thereby control the masses; and furthermore, by affecting the spaces we inhabit, can manipulate and reshape our selfhood: "who controls the past … control the future; who controls the present controls the past" (Orwell 1948, p. 37).

Orwell's now recognizable environment is pertinent to this collection of essays on the nature of surveillance; from the manner in which spaces can affect identity; to how the gaze of these technologies can determine individual behaviour and selfhood. Jeremy Bentham's panopticon (1843) inculcated surveillance within architecture. What Bentham ascertained was the creation of a consciousness solely based on permanent visibility as a form of power; in effect, a space "based on a system of permanent registration" (Foucault 1975, p. 196). Orwell's urban landscape is not dissimilar in its structure; where "you had to live—did live from habit that became instinct—in the assumption that every sound you made was overheard, and except in darkness, every movement scrutinized" (Orwell 1948, p. 3). Big Brother exhibits the same power as Bentham's panopticon, constantly observing the bodies of Oceania; only in Orwell's version, the power of surveillant technologies not only affects behaviour through a system of power, it also creates identity, where "each individual is fixed in his place … the gaze is alert everywhere" (Foucault 1975, p. 195).

Unlike the inmates of Bentham's prisons, Winston Smith is fully aware of the potential control wielded over him by the all seeing eye of Big Brother, and rather than behave, he merely performs correctly: "he had set his features into the expression of quiet optimism which it was advisable to wear when facing the telescreen" (Orwell 1948, p. 5). Winston, under the gaze of the telescreens, continually shifts his identity in order to reflect a visibly acceptable, and more importantly, conformist identity. What we witness with Orwell's form of surveillance is not only panopticism, but also how the gaze of surveillant technologies can shift identity within spaces of visibility. However, as readers of *Nineteen Eighty-Four* will well know, this isn't a system which can be overcome—it is merely a space of continual identity immobility where acts of individualism are punished, and a life of perpetual performativity upheld. As Michel Foucault's work on panopticism states: "power has its principle not so much in person as in certain concerted distribution of bodies, surfaces, lights, gazes" (Foucault 1975, p. 202), and for this reason, Winston Smith will never be able to defeat the power of surveillance.

What we can learn from Orwell's vision is the manner in which technology can impact our understanding of reality, and with that, the ensuing issues of who is surveilled, who has the power of the gaze, and how architectural structures can impact our surveilled potential. Martin Fuglesang and Bent Meier Sorenson's work on Gilles Deleuze extends this idea further, where identity is marked by spaces and given corporeality by its action (Fuglesang and Sorenson 2006). In *Nineteen Eighty-Four*, Winston's identity is almost entirely created by the surveillance of Big Brother and by Winston's actions—he is both given identity as a body to be watched, and one to be watched as he disobeys the laws of Oceania. However, according to Fuglesang and Sorenson, if we require a frame in order to have an identity (in Winston's case, his apartment and the telescreen) then it is, rather confusingly, this same frame which allows us to be real—in essence, it is by being watched that we can become real. The potential contained in surveillant technologies is therefore twofold: providing bodies with identities they may not want, but at the same time providing them with an identity that can be determined as real – I am watched, therefore I am. Surveillance technologies may not, therefore, deserve their dystopian image, and as these chapters suggest, may contain the potential for individuals to become more than just a body to be watched.

Orwell's Oceania is arguably, the most recognisable fictional example of a surveilled state, but the reality of surveillance in the modern world appears to be far more entrenched than the all seeing eye of Big Brother. Salient media discourses remind us that surveillance exists all around us, and in a multitude of forms. In May 2016, artist Laura Poitras exhibited "Astro Noise" at the Whitney owned Hurst Family Gallery in New York. The exhibition consisted of a series of documentary clips, architectural plans and documents, and thermal radiation images on the subject of mass surveillance and the US drone project. Hailed as a form of "political art" which "reveals mass surveillance at home and [the] extensive drone wars abroad" (Cotter 2016a, p. C21), the exhibition exposed Poitras' involvement with the Edward Snowden files following her collaboration with the documentary film *Citizenfour*. Some of the exhibition was shaped by the Snowden leaks and featured images of rooftops in Baghdad and slow motion images of New Yorkers staring at Ground Zero. One particular exhibit, "O Say Can You See" featured a two sided video installation of black and white footage of prisoners in Afghanistan cut with the military aftermath of 9/11. As Holland Cotter determines, the exhibit draws attention to the need for survival in the age of mass surveillance, calling it "art of the 'we shall overcome' sort" (Cotter 2016b, p. C21). Poitras' exhibition unveils the impact of surveillance not only on ourselves, but also on our understanding of others. In a similar vein, in June 2016, the International Center of Photography (ICP) in New York unveiled its inaugural exhibition entitled "Public, Private, Secret". The ICP's curator, Charlotte Cotton, conceived of the exhibition in an attempt to address timely questions concerning how the images we broadcast communicate something about ourselves, and what happens when others view these images without our knowledge (Budick 2016). Much like Poitras' exhibition of post 9/11 surveillance, Cotton's exhibition confronts us with images from webcam stills, Instagram and Twitter in a manner that renegotiates the links between self, viewer and other (Budick 2016). These exhibitions demonstrate, not only the prevalence of surveillance technologies, where real surveillance can be turned into art, but also the many forms surveillant practices can take where, "visual fiction stands in for truth" (Cotter 2016a, p. C21). This collection returns both to 9/11 and to the promise of art in order to capture our current surveillant reality.

Perhaps the most recognizable images of surveillance are those found on the Internet. Much like the ICP's exhibition, the Internet and, by extension, web imagery found on social media, has quickly become a form of individual surveillance—observing ourselves through 'selfies' and others through 'likes'. In 2009, Ondi Timoner made a documentary about Internet pioneer Josh Harris entitled *We Live in Public*. Timoner's documentary charted the rise and fall of Harris' career from the dot com boom of the late 1990s, to his meteoric fall following the art project 'Quiet'. In 1999, Harris invited one hundred artists to live in a human terrarium under New York City where their every move was followed by cameras. Living in "pods" these artists could tune into other people's monitors around them, viewing each other's CCTV channels and living constantly "in public". Harris' experiment took the loss of privacy in the Internet age to a new level, resulting in participants claiming to feel like "rats" and "slaves" and reporting a loss of identity and increasing detachment. As Harris' opening lines in Timoner's documentary disclose, "the Internet is like a new human experience. At first everyone's going to like it, but there will be a fundamental change in the human condition. As time goes by we will become more constrained in these human boxes. Our every action will become accountable. One day we will wake and realise we are all just servants" (*We Live in Public* 2009). Harris' words strike a chord with the central concerns contained in this collection—the manner in which states, selfhood and indeed spaces can be affected by our constant watching and being watched; where, according to the film maker, we will become nothing more than a zoo where we ourselves are the animals (Timoner 2009). Both Timoner and Harris later reflected that we are fast approaching this state, where we are already "trapped in virtual boxes" through our Blackberry and iPhone devices (Alter 2009). As Timoner's film, and Harris' project make clear, Orwell's Big Brother is no longer an individual watching our every move, rather it is a collective consciousness maintained and perpetuated by our need to feel constantly connected through technology, where our engagement with contemporary technologies alters the multiple sets of social relations in which we exist.

Many of these examples touch on a central issue at the heart of surveillance studies—privacy. Orwell's Big Brother denied privacy; Poitras' exhibition demonstrated its loss; the ICP's exhibition unveiled our imaging obsessions; and Harris' 'Quiet' demonstrated our inability to be truly private in a world of growing communication. Perhaps nowhere is this more evident than in our investment in social media where our personal video and imagery share the same tenets as news footage and global surveillance technologies. Twitter, for example, has revolutionized the way news reports are written and broadcast, becoming increasingly integrated into the realm of public statements. We now consume and produce content, becoming what Toffler originally termed "prosumers", embroiled in a culture of speed and sound bites. As one journalist has noted, the manner in which Twitter engages with news events as they unfold, ultimately threatens to undermine more than just the integrity of the news they report, where "embedding a third party hosted tweet in a news article has consequences for privacy—it gives Twitter the possibility of tracking the reading habits of users around the web" (Higgins 2015, [online]). The ability for our actions to be tracked, monitored and surveilled through the Internet is not dissimilar to Harris' vision of our future, where a "Wired City cyber ship" will host and observe our every move, rewarding us for "likes" with cash in place of a "normal" job (Boys-Myer 2011). Many television series and films have similarly warned us of the dangers of losing our privacy through surveillant practices such as CBS' *Person of Interest*'s 'Machine' (2011–2016), *V for Vendetta*'s street spy vans (2005), *Minority Report*'s retina scanning (2002), and *The Truman Show*'s pastiche of reality television (1998). *The Truman Show*'s inversion of reality television (seen elsewhere under the guise of entertainment in shows such as *Big Brother* (2000—present), *Keeping up with the Kardashians* (2007—present) and *Geordie Shore* (2011—present)) has a much darker side to its pastiche of television and privacy. Peter Weir's film also includes shots of the audience who scrutinise Truman's every move in a similar fashion to ourselves as we watch the film. Watching *The Truman Show* is perhaps more uncomfortable viewing than the Kardashians as we suddenly become aware of our own part to play in the surveillance structure; we too are his audience, maintaining his position as entertainment. Given that Truman's identity is also mixed with advertising, the play on modern television is further compounded, reminding us that information, whether it be real of artificial can always be fashioned into an identity (Marks 2005).

Yet these artistic and fictional accounts reflect a much wider concern regarding the impact of surveillance. The events that unfolded during 9/11, Hurricane Katrina and the much broader Western 'war against terror' all share a common denominator with the above examples—the filming and surveilling, as well as the viewing of individuals. Much like *The Truman Show* there is an implicit voyeurism attached to the televised viewing of these disasters. As Laura Mulvey's work on scopophilia details, the idea of being looked at and the pleasure derived from gazing at a screen, is implicit in the viewing of surveillance footage—a form of gazing that connects identity on screen to that of the viewing individual (Mulvey 1973). During 9/11, CNN ran live footage of the disaster and the imagery shot by news crews was reborn in numerous documentaries, such as Jules and Gedeon Naudet's *9/11* and the *Here is New York* exhibition that featured a collection of photographs gorily entitled "*Victims*". Perhaps the most iconic image of 9/11 is Richard Drew's photograph of *Falling Man*, an image which still manages to encapsulate the events of 9/11 despite its unnamed subject. Arguably, it is these images that remain in the public consciousness, overtaking the event itself to become the defining image of all that 9/11 was and became, deriving an identity, of and in itself, as the "real" and authentic version of events. As Junod's article in *Esquire* suggests, *Falling Man* is now "falling through the vast space of memory" (Junod 2003, [online]). By extension, the surveillance technologies throughout the Middle East, and specifically in Afghanistan in order to combat the 'war on terror', have become part of the modern landscape. Much like Orwell's vision, in Kabul, such spaces of continual surveillance are common place where "the blimp has become their constant overseer … 'It watches us day and night'" (Bowley 2012, p. A15). The impact of this is felt in our understanding of the modern world, where our vision of the Western war on terror is almost entirely derived, and indeed, contrived, by the all-seeing eyes of drones. Much like the imagery of 9/11, and Poitras' exhibition, we consume camera footage through our television screens in news reports, and in imagery on the Internet in order to observe and ultimately feel connected to the world around us.

Surveillance therefore, features much more in the shaping of our identity than it may first appear to. As these examples demonstrate, our sense of identity can be formed by the spaces of surveillance, the images of surveillance, and even the act of looking itself. And we ourselves are part of the expansive documenting of bodies, where, beyond the observing of others, we too are part of the surveillance society. Consider the use of wiretapping and speech recognition software, (long outdated since the arrival of the Internet in 1991), which is now replaced by government databases and pattern recognition software which cross-correlates our Internet surfing against our likes and dislikes, often "stealing" information from social media to personalise advertisements. In 2012, both Google and UK mobile phone networks collected data from their users through the collation of users' email content and web traffic (Cellan-James 2012), and web pages are increasingly utilizing various data collection methods such as: Cookies (when the website "knows" you have visited before), Clickstream Data (recording where the mouse clicks), Search Data (based on frequently searched terms), Purchase Data (online shopping) and Profile Data (from social media). Financial transactions can now be tracked, our vehicle number plates 'recognised' and our mobile phones tracked through GPS (Global Positioning System). Whilst it may have once been relegated to the realm of science fiction, DNA profiling, facial recognition and body scanners are all part of our modern culture, indeed, part of the societal systems which aim to ease our concerns over our identity and our possessions. So expansive is surveillance, that a quick search on the Internet for the terms "Mass Surveillance" brings hits for: 2013 Global Surveillance Disclosures; Computer and Network Surveillance; Data Privacy; Data Retention; Government Databases; National Security; Pen Register; Phone Surveillance; Police State; Security Culture; Sousveillance; and Tracking Systems. Far from being something we witness in art exhibitions, dystopic novels and television programmes, surveillance occurs on our computers, through our phones, in our homes and during our conversations.

Mass surveillance pervades our daily lives, but with CCTV cameras and securitynetworks comes a reactionary phenomenon—the rise of personal surveillance. In 1990, Cop Watch sought to reverse the surveillance culture in America, by filming police officers who stopped and searched individuals. The movement gained notoriety in 2003 when Kendra James was fatally shot in Portland as she drove away from a traffic officer, and Cop Watch has continued to re-use surveillance technologies against police agencies in an attempt to take back some form of bodily control (Monahan 2006). More recently, the movement gained followers from the Black Lives Matter movement—a movement which emerged from the Ferguson riots and the killing of Michael Brown in 2014. Black Lives Matter has itself been the subject of surveillance, as the *Intercept* claims "the department [of Homeland Security] frequently collects information including location data, on Black Lives Matter activities from public social media accounts, including Facebook, Twitter and Vine" (Joseph 2015, [online]). Personal surveillance seems to be growing in response to the global surveillance at the core of governing bodies, where even cyclists in London have installed GoPros on their helmets in an attempt to offer security against unsafe drivers. The correlation between seeing and controlling is apparent – a symptom of the insecurity of the individual.

It is clear surveillance is a far reaching, and immeasurably myriad subject, affecting spaces, selves and states in profound and sometimes invisible ways. As the examples considered in this Introduction demonstrate, surveillance is often capable of either reducing our identities, but also, occasionally, enhancing them. As David Lyon's seminal works on surveillance culture testify: "thinking about surveillance today ... means confronting the broken, stranded presents of a temporal disunity, shining with an eerie glossiness ... meaning is an effect produced in the passage from signifier to signifier" (Lyon 2006, p. 127). In a manner similar to the earlier consideration of Deleuze, there is a complex issue at the heart of these technologies which questions our selfhood—by being watched are we made real? As Lyon makes clear, effect can only be produced when there is a passage between signifier to another signifier. By being viewed, do we pass over some signification from and to the camera, each enabling each other's existence and thereby making each other real? Recalling the works of Frederic Jameson, in this postmodern world of constant observation, there is some truth behind the idea that through camera intensification, we can become "more literal and more vivid" (Lyon 2006, p. 126); but this intensification comes at a cost—where we are instead "disconnected from the other signifiers that give ... an identity" (Lyon 2006, p. 129). Much like Timoner's documentary, Harris' "Quiet" proved exactly this double bind—we are at once desperate for connection, but we seek it by watching others, who in turn watch us, distancing us from the real; a cycle that can never be diminished or satiated. William Staples' *Everyday Surveillance*

(2014) similarly reflects on modern society by suggesting that the collection of data has always existed, whether through technological means or otherwise (Staples 2014). For Staples, technology is frequently employed to 'watch' the body—its habits, movements, activities and, ultimately to shape, or change, its behaviour so that identity, becomes data. What Staples makes clear is that we are entering into a metaphorical age of Bentham's panopticism where we will be permanently visible and susceptible to the technological gaze. Rapheal Sassower's *Digital Exposure* (2013) argues that far from standardising the capitalist market, the introduction of consumer surveillance has resulted in the personalisation of consumption (Sassower 2013). By collecting data on individuals, there is no longer a drive to create uniform consumption but rather, individualised purchasing power, reversing the Orwellian idea of surveillance as a malignant force in modern society. Shoshanna Zuboff's *In the Age of the Smart Machine* (1988) similarly expresses the idea that technology can be understood as a duality, and not a simplistic and negative entity. On the one hand, when applied to automatic operations that can replace the human body, technology is seemingly humanity's eradicator (Zuboff 1988). On the other hand, technology generates information about production and human and bodily action. Activities, events, and objects are translated when a technology informates as well as automates. Technology can therefore, be both the eraser of humanity and inform on it, and for this reason, remains misunderstood and feared, entering into the double bind expressed by Lyon. Arguably, Michel Foucault's work on panopticism in *Discipline and Punish* (1975) offers the most established criticism on surveillance studies. For Foucault, visibility is a trap—by being visible, we ensure power can be exerted over our bodies (Foucault 1975), and whilst we may not require a panopticon in every street, its methods have been adopted by modern society through systems of control and observation. As these writers make clear, surveillance almost certainly affects our understanding of individual liberty and identity. This edited collection seeks to interrogate some of these established discourses through an examination of cultural productions, expanding on these established discourses and into new interdisciplinary discussions. The three sections included here demonstrate how art, photography, film, literature and place can impact the body, investigating issues such as how trauma affects how bodies are viewed; how sight, and vision can be associated with othering; how identity can be complicated by the 'camera eye'; how technology can mutate spaces into panopticons; the effects of clinical development; and how the body is policed in a post 9/11 climate.

Unlike previous studies, this collection aims to develop the discourse of the surveilled self and expand on the lived realities of post-panopticism by unpacking a unique variety of cultural products which articulate contemporary experiences of surveillance. The authors in this collection approach modern surveillance through art, photography, literature and through the body itself, examining how contemporary cultural production offers new insights into the Foucauldian version of biopower which modern surveillant assemblages inculcate. These cultural products repeatedly challenge modern moralities and modes of being, performing or causing us to perform what Bauman and Lyon (2013) terms 'adiaphorization', in which our human processes and systems become disengaged from morality. Bauman and Lyons' term 'liquid surveillance', addresses the manner in which contemporary surveillance is engaged in the colonizing of new and disparate areas of life. *Spaces of Surveillance* develops the vision of the 'liquid' sphere which seeps into each area of modern life and into our performance of selfhood at each and every level; into art, literature, film, lived spaces, and psychic worlds. This collection engages with the cultural aspects or the "felt" elements of this global discourse. *Spaces of Surveillance* contributes to these ongoing debates about privacy, selfhood and visibility, but does so from a position of cultural expansiveness, engaging with established discourses on 9/11, government surveillance and the collation of personal information through a collection of chapters which examine the repercussions of these desires, both for the individual and for society. This edited collection attempts to consider how spaces, places and states of selfhood have been impacted by surveillant technologies, offering a unique insight into the ways in which bodies have both voluntarily and involuntarily been shaped and defined by changing technology. The exponential growth of these technologies purports to facilitate our self-management, security and control, and the ubiquitous character of contemporary 'selves'. Entry and exit from multiple identities and networks is seemingly made easier, but as the following chapters illustrate, the psyche is not always comfortable in these liquid surveillant times.

This collection features ground-breaking analyses of a variety of approaches to, and perspectives on, contemporary surveillance as it is expressed in cultural production. The first section in this collection, "Art, Photography and Film" features contributions which examine the works of artists Jill Magid and Susan Collins. Amy Christmas' chapter engages in the representation of surveillance in art; something which she deems has been popularly seen as a reaction to notions of modern anxieties. While many surveillance art projects—especially those by female artists, such as Sophie Calle and Mona Hatoum—have proved incisive and provocative in terms of their delineations of the contemporary subject, they have in common an entrenched pessimism regarding the state of identity in relation to the state of surveillance. This chapter explores the early work of American multimedia artist Jill Magid and investigates several artworks and writings from 1999–2004. Christmas' chapter considers the 'lack' observed within the field of surveillance studies, working with, in Magid's words, "function at a distance, with a wide-angle perspective, equalizing everyone and erasing the individual" (Magid 2000). In an attempt to harness the potential of these technologies to exploit a way to 'see', and thereby construct selfhood in our own terms, Christmas explores the development of Magid's subject through the exploration of the body of the artist. Examining works such as *Kissmask, Surveillance Shoe, Monitoring Desire* and *Evidence Locker* this chapter explores how the writings of Magid cohere with the installation pieces, and by extension, how they contribute to surveillance studies through engagement with the individual's experience of being observed. The second chapter on Susan Collins' *Glenlandia* by Jaclyn Meloche, similarly questions the impact of surveillance on the individual. In *Entangled: Technology and the Transformation of Performance* (2010), Christopher Salter discusses the transformative function of the camera in order to argue that its frame is no longer a device, but rather is a mechanical lens which can reproduce meaning (Salter 2010). Meloche outlines this same theoretical framework of performativity and digital performance in order to complicate the function of the webcam and mechanical lens in digital art, examining the power of the gaze in Susan Collins' work. This chapter investigates the questioning of representation in lens culture, and by extension, the performativity implicit in digital surveillance of the body. By reinterpreting Collins' work, Meloche's chapter modifies the technological function of the webcam into one of selfhood and identity formation. By becoming spectacle as Meloche suggests, Bill McBride's chapter on the film *Her* (2013) extends the debate about shifting bodies and selves into new and perhaps, less autonomous versions of identity. McBride examines the possibility for the flow of subjectivity to be reversed, where machinery becomes threatening by posing the potential to become more human than ourselves. Set in the near future, Spike Jonze's film explores the evolving nature of love with an operating system called Samantha. Through an investigation of ethical dimensions of artificial intelligence and cybernetic threats to humanity, McBride analyses the film's stylized gestures to recover and define what it means to be human. What emerges from this chapter is the potential for bodies to mirror their surroundings—in effect, to become machine-like, once again questioning the notion of realness in a world of surveillance technologies and increasing mechanization. In continuation of the pursuit of "realness", Frances Pheasant Kelly's chapter examines scopic strategies for monitoring criminal transgression through an examination of Kathryn Bigelow's *Zero Dark Thirty* (2012). Pheasant Kelly determines that in a similar vein to films made after 9/11, Bigelow's text promotes American notions of surveillance for the greater good and protection of democracy, but does so from a position which justifies the intrusion of the spectator into privacy. Following the observation of Bin Laden's location through US military technologies, Pheasant Kelly observes that the film's narrative corresponds with Foucauldian theory for almost its entirety before making Bin Laden visible in the final scenes. Engaging with the work of Mathisen (1997), this chapter considers the multiple scopic regimes evident in the film, and pays attention to the relationship between surveillance and terrorism in the context of post 9/11 America. The final chapter in this section focuses on trauma, extending Pheasant Kelly's investigation of the trauma of war, Simon Bacon's chapter questions the manner in which trauma can impact bodies and their identity. Referring to Pramod K Nayar *Citizenship and Identity in the Age of Surveillance* (2015), Bacon considers the manner in which surveillance links directly into the formation of the global subject and how cultural trauma can construct an identity, where trauma is a loss of the collective, national, global identity and that the personal one created in front of the camera can become unreal, or even unnatural. This unique chapter argues that the vampire is a symbol for national and cultural disconnection, which goes beyond not only humanity, but also national belonging; arguing for the victims of the vampiric gaze to have a constructed identity formed entirely from trauma. Examining films such as *Vampires* (2010), *Afflicted* (2013), and *Daybreakers* (2009), Bacon's chapter extends the discussion of the surveilled subjects from earlier chapters, and reconfigures them into those who have been infected by the camera, resulting in a self-surveilling selfhood.

Bacon's reading of film, in line with Pheasant Kelly's, questions an identity that appears "Othered" by surveillance—a distancing which can render the body as unreal and alien by its position on screen. This is further developed in the second section of the collection through an examination of surveillance in literature. Alison Lutton's chapter on Bret Easton Ellis' *The Informers* (1994), considers the spaces of Ellis' narratives as those of observation and constant gazing. With brief consideration of *American Psycho* (1991) and *Less than Zero* (1985), Lutton's chapter investigates the postmodern urban environment as one of a cultural and spatial framework premised upon observation of varied contested locales, as featured in *The Informers*. Her chapter unveils the plurality of spaces and the overlooked surveilling subjects therein, foregrounding contemporary cityscapes of Los Angeles and Tokyo as representative of 'feedback' spaces, thereby continuing the discussion of identity placement and how the act of surveillance can suggest a performativity of bodies within its gaze. Caleb Milligan's chapter on Don DeLillo's *White Noise* (1984), argues that reading DeLillo's text alongside the emerging technologies of that year, can help us to better understand how the "surveillance of sousveillance" can be traced into our current cultural moment. Citing the surveillance prophecy of Orwell's text, Milligan discusses the impact of monitoring oneself as much as we too are monitored in modern society, and in doing so, analyses the fear of, and desire for, the computer's coded messages in *White Noise* to contribute to current culture. Contending the development of gaming as a medium to become 'other', we too seek to become extended selves through surveillance technologies. Rather than categorize surveillance as only "them watching us," he argues it is "us watching each other"; a clarification that nuances the idea that the surveillance of sousveillance constitutes our trend of being watchable to give others something to watch. Therefore, through specific applications of literary and media studies, this chapter contends that our cultural narratives are affected by not merely the surveillant gazes of multiple media, but our sousveillant desire to connect with those gazes. This connection with multiple gazes established in Milligan's chapter, is extended by the discussion of the ethical complications present in digital narratives through Virginia Pignagnoli's chapter on Dave Eggers and Jonathan Franzen. Pignagnoli's chapter provides a unique and highly original analysis of the narratives created in the Internet age, investigating the relationship between surveillance, technology and literature and the ethical issues they raise through their encounter with one another. Examining three recent novels, *The Circle* (2013), *Purity* (2015) and *Super Sad True Love Story* (2010), Pignagnoli's chapter explores narratives produced in the Twenty-First century, and the main rhetorical resources the authors employ to express a world of poor communication, shallow selfhood and the digitization of our lives. Through the mapping of narrative spaces contained in these novels, and comparison with geographical spaces, Pignagnoli determines the creation of a totalitarian space entirely determined by digital communication, the result of which is a world where surveillance becomes both the principal concern of the narratives' ethical dimension and the framework through which the novels become an implicit plea for the printed book against the rise of digital technologies. The urgency of these troubled communication forms is both implicit and explicit: the chapter urges action while it acknowledges the contradictions in which we now live. The final chapter in this section from Jeffrey Clapp, on selected works from Claudia Rankine, investigates the notion of ethics and communication in literature. Clapp's exploration of Rankine's *Citizen* (2014) and *Don't Let Me Be Lonely* (2004), considers the stakes of surveillance, counter surveillance, sympathetic identification, and media culture in the post 9/11 period through an examination of citizenship and the position of the spectatorial subject. This remarkable contribution to the collection addresses the radicalization of the self, post 9/11, proposing that Rankine's pronoun work can be understood as a reflection on the strange proximity of surveillance society to the society of the spectacle, confusing classically liberal demands for political representation and personal expression.

The concluding section of the collection entitled "States, Place and Bodies" engages with notions of the bodily self as it manoeuvres through the complex systems of contemporary surveillance. The section examines the use of surveillance in the practice of 'profiling' and interrogates the implicit quest for an essence of selves. Sam Tecle, Yafet Tewolde, Tapo Chimbganda and Francesco D'Amico's chapter investigates, in the words of Simone Browne, blackness as a metaphor, in order to reveal how surveillance has been constituted in, and through, antiblackness in Canada. Employing Browne's framework, the chapter considers the multiculturalism of Canada's national integration policy, exploring practices, performances and policies in order to demonstrate how the surveillance of blackness facilitates and results in the racial profiling and management of blackness that is already an historically ordained issue. Through an extension of the concept of racial management, the authors of this chapter suggest that popular culture, higher education and sites of activism within the Canadian context, are all spaces of surveillance, and begins to address issues of race in relation to national borders. Mary Ryan's contribution addresses the concept of sousveillance, as it identifies cultural trends and political implications through political and social agendas. Engaging in issues of body worn cameras, Ryan's chapter examines a broad range of participants, from individual actors to government officials, in order to reflect on the implications of sousveillance and modern surveillance systems in civic engagement, political discourse, and socio-political experiences in American life; ultimately evaluating the threats, as well as opportunities these practices offer. This chapter engages with some positive possibilities of this new sous-scopic regime. Susan Flynn continues this theme to some extent, attempting to identify positive angles as she investigates medical surveillance through a consideration of digital imaging, diagnostic techniques and biometrics. In this chapter, she explores the manner in which emergent medical techniques bring new forms of knowledge that can shape the understanding of ourselves. By critiquing that information which is omitted from the data we share, this chapter attempts to identify the parameters of the medical gaze. Flynn's chapter considers the configuration of bodies, and by extension identity, through surveillance technologies both as positive (through a sense of possibility) and negative (the interruption of the body's privacy) through regimes such as data sharing. The chapter attempts to reconceptualise sousveillance as both a means of sustaining biological integrity and maintaining individual intellectual freedom, as the graph leaves our inner individuality to our own control.

The afterword to the collection is provided by Professor Vian Bakir, who considers the authors' contributions in light of the recent Snowden leaks. Structured in two parts, Professor Bakir's afterword recognises the rising tide of surveillance awareness in both critical and practical scholarship, noting the depth and pervasiveness of what and whom is surveilled in corporate and digital forms of surveillance. The second section considers the impact of public opinion polls on surveillance and privacy, identifying the need for greater digital literacy among the public in order to make sense of the scale of the surveillance culture unveiled by the Snowden files. Ultimately, Professor Bakir's afterword highlights the original contribution this collection makes in the humanities, asking what does it feel like to be surveilled, and, perhaps more importantly, does it really matter?

This book seeks to bring a multitude of surveillance based discussions into dialogue with each other, questioning why, when, and by whom, are we surveilled. *Spaces of Surveillance* questions the subsequent formation of identity and states of selfhood from within a cultural context where surveillance is not only around us, it is already part of who we are. If we are already within the surveillance matrix, then the question surely isn't will we always be watched, but rather what will happen to us when we are watched; and furthermore, if our bodies are agents of these spaces, will we ever be truly free? This book examines the idea of being watched, of policing ourselves and others, through the notions (both literal and theoretical) of surveillance. In doing so, these chapters begin to question our understanding of privacy, how identity can become complicated by technology, and what this intrusion can mean in terms of our being 'real' in a surveilled world. Television programming and Hollywood movies, continue to exploit our fears over being watched in nightmarish versions of surveillance systems: in *Homeland* (2011—present), *CSI* (2000—present), *CSI:Cyber* (2015–2016), *Hunted* (2015—present) and movies such as *Gattaca* (1997), *Panic Room* (2002), *The Purge* (2013) and *Cache* (2006). Authors too, continue to write dystopian visions of our technological world: in Margaret Atwood's *The Handmaid's Tale* (1985), Lisbeth Salander's *The Girl with the Dragon Tattoo* (2005) and Suzanne Collins' *The Hunger Games* (2008). What we can learn from this collection, and indeed from the continued interest in surveillance expressed through literature, art, film and television, is that despite the deluge of cultural material concerning our paranoid thoughts—those of someone watching us wherever we go—our paranoia is futile; the matrix is already internalized.

REFERENCES

Alter, A. (2009). Culture—Just Asking/Josh Harris and Ondi Timoner: The Truman show for everyone—A documentary filmmaker and her now reluctant subject on living in public. *The Wall Street Journal*. Retrieved August 14, 2016, from http://www.wsj.com/articles/SB123879482878387933.

Bauman, Z., & Lyon, D. (2013). *Liquid surveillance*. Cambridge: Polity.

Bentham, J. (1843). *The Works of Jeremy Bentham*. Retrieved August 14, 2016, from http://oll.libertyfund.org/titles/bentham-works-of-jeremy-bentham-11-vols.

Bowley, G. (2012). Spy balloons become part of the Afghanistan landscape. *The New York Times*. Retrieved August 13, 2016, from http://www.nytimes.com/2012/05/13/world/asia/in-afghanistan-spy-balloons-now-part-of-landscape.html.

Boys-Myers, C. (2011). The future according to Josh Harris. *Entrepreneur*. Retrieved August 14, 2016, from http://thenextweb.com/entrepreneur/2011/07/24/the-future-according-to-josh-harris-but-wait-whos-josh-harris/#gref.

Budick, A. (2016). Public, Private, Secret: International centre of photography. New York: Meagre. *Financial Times*. Retrieved August 13, 2016, from http://www.ft.com/cms/s/0/57a1b88e-4213-11e6-9b66-0712b3873ae1.html.

Cellan-James, R. (2012). Web surveillance—Who's got your data? *BBC News*. Retrieved August 14, 2016, from http://www.bbc.co.uk/news/technology-17586605.

Cotter, H. (2016a). Laura Poitras: Astro Noise' examines surveillance and the new normal. *New York Times*. Retrieved August 13, 2016, from http://www.nytimes.com/2016/02/05/arts/design/laura-poitras-astro-noise-examines-surveillance-and-the-new-normal.html.

Cotter, H. (2016b). Photography's shifting identity in an insta world. *New York Times*. Retrieved August 13, 2016, from http://www.nytimes.com/2016/06/24/arts/design/review-photographys-shifting-identity-in-an-insta-world.html.

Foucault, M. (1975). *Discipline and punish*. Paris: Edition Gallimard.

Fuglesang, M., & Sorenson, B. M. (2006). *Deleuze and the social*. Edinburgh: Edinburgh University Press.

Heir, S., & Greenberg, J. (2007). *The surveillance studies reader*. Maidenhead: Open University Press.

Higgins, P. (2015). Twitter axes accountability projects. *Electronic Frontier Foundation*. Retrieved August 13, 2016, from https://www.eff.org/deeplinks/2015/08/twitter-axes-accountability-projects-sparing-politicians-embarrassment.

Joseph, G. (2015). Feds regularly monitored black lives matter since ferguson. *The Intercept*. Retrieved August 14, 2016, from https://theintercept.com/2015/07/24/documents-show-department-homeland-security-monitoring-black-lives-matter-since-ferguson/.

Junod, T. (2003). Falling Man. *Esquire*. Retrieved August 14, 2016, from http://classic.esquire.com/the-falling-man/.

Lyon, D. (2006). *Theorising surveillance: The panopticon and beyond*. New York: Routledge.

Magid, J. S. (2000). *Monitoring desire*. M.Sc., Massachusetts Institute of Technology. Accessed 22 Aug 2011.

Marks, P. (2005). Utopian visions and surveillance studies. *Surveillance and Society*, *3*, pp. 222–239.

Monahan, T. (2006). *Surveillance and security: Technologised politics and power in everyday life*. Abingdon: Routledge.

Mulvey, L. (1973). *Visual pleasure and narrative cinema* (pp. 803–816). Film: Psychology, Society and Ideology.

Orwell, G. (1948). *Nineteen eighty-four*. London: Penguin.

Salter, C. (2010). *Entangled: Technology and the transformation of performance*. Cambridge: MIT Press.

Sassower, R. (2013). *Digital exposure: Postmodernism postcapitalism*. New York: Palgrave Pivot.

Staples, W. (2014). *Everyday surveillance: Vigilance and visibility in postmodern life*. New York: Rowman and Littlefield.

Timoner, O. (2009). *We Live in Public*. [DVD].

Zuboff, S. (1988). *In the age of the smart machine: The future of work and power*. USA: Basic Books.

AUTHOR BIOGRAPHIES

Susan Flynn is a lecturer at the University of the Arts, London where she specialises in contemporary media culture, digital and body theory and media equality. Her work is featured in a number of international collections and journals such as *American, British and Canadian Studies Journal* and *Ethos: A Digital Review of the Arts, Humanities* and *Public Ethics*.

Antonia Mackay is an Associate Lecturer at Oxford Brookes University and Visiting Lecturer at Goldsmiths University of London. She has taught on a wide range of undergraduate and postgraduate modules including: American Theatre, American Vistas, Critical Theory, Narrative and Narratology, Special Options in Experimental Avant Gardes, Contemporary Literature and Twentieth Century Literature. She has published articles on Manhattan Maleness and Cold War ideology, as well as those on space, technology and identity. Antonia was the winner of the 2014 and 2016 Nigel Messenger Teaching Award at Oxford Brookes.

PART I

Art, Photography and Film

CHAPTER 2

Equality and Erasure: Responses to Subject Negation in the Art of Jill Magid

Amy Christmas

Representations of, and engagements with, surveillance in art and literature have exhibited trends in keeping with what Kirstie Ball and Kevin D. Haggerty identify as a dystopian current in the scholarship of the field, when they describe academics and researchers across a range of disciplinary orientations as having "embrace[d] a metanarrative of ever more surveillance becoming more discriminating and intrusive" (2005, p. 136). Meditations on surveillance by writers and artists have tended to articulate the anxieties of popular culture: from Orwell and Zamyatin to Le Carré and the twentieth-century rise of the spy novel to China Miéville's *The City and the City* (2009) and even E. L. James's *Fifty Shades of Grey* (2011)—surveillance has been inextricably bound up with the delineation of power relations and the dialectics of agency that underpin them. The contemporary art world, too, has provided rich commentary and conceptual mapping of the modern surveillance state, from the *noir*-esque performance pieces of Sophie Calle (*Suite Venitienne*, 1980; *The Shadow*, 1981) to the iconoclastic street art of Banksy (*What Are You Looking At?*, 2004; *One Nation Under CCTV*, 2008) to more recent developments in the "artveillance" (Brighenti 2010)

A. Christmas (✉)
Qatar University, Doha, Qatar
e-mail: achristmas@qu.edu.qa

© The Author(s) 2017
S. Flynn and A. Mackay (eds.), *Spaces of Surveillance*,
DOI 10.1007/978-3-319-49085-4_2

field which traverse a range of media (cf. sousveillant lifelogging in the works of Steve Mann and Hasan Elahi; bio-hacking in Heather Dewey-Hagborg's *Stranger Visions*, 2012–2014; street theatre by the Surveillance Camera Players, 1996–2006). Despite such diverse methods and dispositions, from the playful to the politically-charged, one of the engrained ideas that is persistently reiterated regarding surveillance is our passivity in its face; this no doubt stemming from the Foucauldian figuration of docile bodies (1977, p. 136), which goes some way to explaining our cultural complacency regarding the ubiquity of surveillance as a structural feature of our late modernity (Giddens 1990, p. 59).

Two lasting figurations of surveillance have dominated the trajectory of critical thought on the subject in the late twentieth and early twenty-first centuries: George Orwell's Big Brother (*Nineteen Eighty-Four*, 1949) and Michel Foucault's analysis of Jeremy Bentham's panopticon (*Discipline and Punish*, 1975), both of which have encouraged the propensity to perceive the spaces of surveillance as essentially oppressive and deindividuating. Furthermore, their frequent reiteration in contemporary cultural production has provided both popular and academic discourses with iconic references through which experiences of surveillance societies can be readily articulated. Orwell, by employing surveillance as the right arm of totalitarianism, firmly situated the modern socio-political subject—embodied in the novel by the privately anarchic civilian Winston Smith—within an "us versus them" dichotomy that continues to resonate sharply with present-day responses to surveillant technologies and practices. Foucault, in his scrutiny of the "panopticism" generated by the social applications of these technologies and practices, initiated a profound discussion—one which continues to underpin most scholarship in the field of surveillance studies today—concerning the effects of institutionalised surveillance upon subjectivity and the nature of the identity work that may be enacted within surveillant spaces. That these two powerful metaphors have gained so much traction in both civic and scholarly discourses implies a consensus that the self is suffering in the surveillance society. As much can be inferred from the conclusions drawn, not only by philosophers, but also the artistic and literary practitioners who, more and more, are choosing to implement surveillant themes in their work. How, then, might we circumvent traditional dichotomies of power that have come to characterise this dystopian metanarrative bolstering the surveillance state, these sketches of passively docile subjects that so greatly impinge upon our constructions of self in the contemporary era?

Jill Magid, an American multimedia artist, works almost exclusively with systems. Her oeuvre to date includes projects concerning institutions ranging from the NYPD and the Dutch Secret Service to criminal forensics teams, the US military and war correspondents. The various institutions depicted in these projects have in common the feature of being closed systems: difficult to penetrate, to come to know. Magid sees the location of points of entry to such systems as an integral component of her artistic practice, which ultimately shapes the technique and tone of the resulting works. Magid's early-career (1999–2004) pieces have experimented with surveillance systems from numerous angles, and while Magid is clearly by no means the first artist to look critically at surveillance and incorporate it thematically into her process, she remains one of the few who have engaged with surveillant technologies and practices in such a way that conventional perspectives are challenged and refreshed, and many positive implications of surveillance emerge. This chapter will take several of Magid's performance pieces and consider the identity work that is taking place within the scopic region of surveillance systems and the institutions they signify. Of her process, Magid writes:

> The systems I choose to work with function at a distance, with a wide-angle perspective, equalizing everyone and erasing the individual. I seek the potential softness and intimacy of their technologies, the fallacy of their omniscient point of view, the ways in which they hold memory (yet often cease to remember), their engrained position in society (the cause of their invisibility), their authority, their apparent intangibility – and, with all of this, their potential reversibility. (2007, p. 1)

David Rosen and Aaron Santesso stress that "any account of surveillance must also consider the ultimate target of all surveillance activity: the individual self" (2013, p. 3). Magid's work with surveillance technologies necessarily interpolates the subject into the system, and in doing so, explores the conditions of the system even as identity is being exercised. As Magid identifies, one of the main functions of surveillance is to forget— those who are remembered are usually those recorded in connection with crime, or perceived antisocial conduct, and the rest are indiscriminately ignored. Surveillance's selective memory negatively constructs identities, while what Robert Knifton has termed the "amnesia of the archive" (2010, p. 93) negates them entirely. Where Magid's work produces a particular cultural resonance is in its capacity to revisualise subjectification in surveillance societies, and, in resisting the equality and erasure of surveillant practices, uncovers the "potential reversibility" of surveillance technologies themselves. This study seeks to investigate the alternative responses present in Magid's thought provoking art in a way that directly engages with the subject's experience of surveillance, rather than surveillance's experience (and dominance) of the individual.

SURVEILLANT ASSEMBLAGES AND DIVIDUAL SUBJECTS

When articulating the effects of surveillance upon subjects, Kevin D. Haggerty and Richard V. Ericson's more recent notion of the *surveillant assemblage* may be added to the established metaphors of Big Brother and the panopticon, which they "draw from a different set of analytical tools" in order to move beyond the limitations of both Orwellian and Foucauldian delineations of surveillance environments, so as to avoid reproducing a "general tendency in the literature to offer more and more examples of total or creeping surveillance" (2000, pp. 607–608). Taking the notion of the "assemblage" as conceptualised by Gilles Deleuze and Félix Guattari (1987)—which describes "a multiplicity of heterogeneous objects, whose unity comes solely from the fact that these items function together, that they 'work' together as a functional entity" (Patton, 1994 cited in Haggerty and Ericson 2000, p. 608)—Haggerty and Ericson propose its application to surveillance culture in order to express the "convergence of what were once discrete surveillance systems" (p. 606). Their Deleuzian surveillant assemblage, they continue:

> … operates by abstracting human bodies from their territorial settings and separating them into a series of discrete flows. These flows are then reassembled into distinct 'data doubles' which can be scrutinized and targeted for intervention. In the process, we are witnessing a rhizomatic levelling of the hierarchy of surveillance, such that groups which were previously exempt from routine surveillance are now increasingly being monitored. (p. 606)

These data doubles are constructed for the specific purposes of peopling particular narratives driven by particular agendas; their abstraction here denotes removal of context, while separation indicates the disassembly of individuals into various and quantifiable categories, the reconstructed sum of which does not necessarily account for all of those parts.

For instance, when considering the ways in which surveillance technologies are implemented to facilitate social profiling, we are reminded that such practices require the observer to categorise and compartmentalise aspects of individuals in order to make control of them manageable. Didier Bigo repurposes Foucault's panopticon as the 'Ban-opticon' to this end, demonstrating how modern policing employs this *dispositif* as a means of social sorting: "A skin colour, an accent, an attitude and one is slotted, extracted from the unmarked masses and, if necessary, evacuated" (2006, p. 46). David Lyon concurs with, and extends, the "panoptic sort" when he notes that surveillance "classifies and categorizes relentlessly, on the basis of various—clear or occluded—criteria. It is often, but not always, accomplished by means of remote networked databases of whose algorithms enable digital discrimination to take place" (2003, p. 8). Within the judicial sphere, the criteria for sorting, surveilling and thus separating an individual may include, but may not be limited to: fingerprints; DNA samples; facial recognition and other biometrics; ethnicity; language/dialect/accent; behaviour; and so on. Such taxonomies of persons are not restricted to the prevention or investigation of criminal activity: with the rise of the Web, surveillance has also flourished in the commercial sector, as Jason Pridmore and Detlev Zwick affirm: "The shift to digitized information [...] is perhaps the most important aspect for understanding the monitoring and measuring of consumers and their consumption practices" (2011, p. 270). In the online marketing sector, individuals are further reduced to specific data flows based on their consumer habits and preferences, customer profiles, even "presumed economic or political value" (Gandy 1993, p. 2). The extent to which social sorting and profiling takes place via the frame of surveillance points to a fundamental shift in how individuals are treated by today's "post-panoptical" (Bauman 2006, p. 11) society, as surveillant assemblages conspire to systematically dismantle identities into manageable, quantifiable, marketable strands of information. The ways that these informational fragments are interpreted, and then restored, in order to create a composite identity—criminal, client, or consumer—are informed by specific agendas and are consequently subject to a high degree of institutional bias.

The phenomenon of the dis- and reassembled subject is also treated by Deleuze when he posits the notion of the individual refigured as *dividual* through its reduction to data flows in the post-panoptic era. In "Postscript on the Societies of Control", he writes that "the numerical language of control is made of codes that mark access to information, or reject it. We no longer find ourselves dealing with the mass/individual pair. Individuals have become 'dividuals', and masses, samples, data, markets, or 'banks'" (1992, p. 6). The Deleuzian outlook for the subject is bleak, and feeds back into the dystopian myth anchoring the surveillance state at both macro- and micro-levels of experience. In spite of this, one can observe contrary depictions of identities presented in surveillant spaces that turn out to be not only deindividuating, but also dividuating.

In their article, Haggerty and Ericson examine the way that the body-as-assemblage "is comprised of myriad component parts and processes which are broken-down for purposes of observation" (2000, p. 613) through the encounter with the surveillant assemblage. Briefly acknowledging the paradigmatic work of Donna Haraway (1991), they recognise that "the monitored body is increasingly a cyborg: a flesh-technology-information amalgam" (2000, p. 611), yet their cursory handling of this critical link between the body-as-assemblage and cyborgian subjectivity overlooks a key opportunity for readdressing identity work in surveillance environments. Haraway's ontological cyborg is as rich a metaphor for postmodern subjectivities as Foucault's panopticon is for the society which impacts upon those subjectivities; by introducing the surveillant assemblage at this juncture, significant conceptual connections can be drawn between the two.

In her "Manifesto for Cyborgs", Haraway insists on the utility of "cyborg imagery" for expressing our postmodern conceptions of techno-cultural self by "taking responsibility for the social relations of science and technology" and "refusing an antiscience metaphysics, a demonology of technology, and [...] embracing the skilful task of reconstructing the boundaries of daily life, in partial connection with others, in communication with all of our parts" (1991, p. 181). Refusing to be disciplined into modes of being by demonized technologies, in fact, refusing to contribute to the demonization of technology at all, the cyborg haunts the peripheries of the surveillant assemblage, mimicking its qualities. The cyborg is "a kind of disassembled and reassembled, postmodern collective and personal self" which recognises that our contemporary identities are "contradictory, partial, and strategic" (pp. 163, 155). This resonance with the subject-as-assemblage—as produced by surveillance practices: the surveillant subject–is striking. The cyborg and the surveillant subject occupy two sides of the same coin, the coin in question being the dividual identity that emerges in opposition to, or perhaps as successor to, individualism. What differentiates Haraway's cyborg from Deleuze's subject-as-assemblage is the way each responds to its own dividual nature: Deleuze sees the dividual subject as a casualty of the control society; Haraway, by contrast, indicates that not only does the cyborg emerge concomitantly with the late-twentieth-century technoscientific shifts in society, it also stands in resistance against "the final imposition of a grid of control on the planet" (1991, p. 154).

Jill Magid invokes the image of the cyborg in her thesis "Monitoring Desire", seizing upon Haraway's configuration of cyborgian subjectivities as "strategic" (Magid 2000, p. 20). Magid's understanding of the way surveillance technologies work to shape, or define, the individual hinges upon two fundamental factors: equality and erasure. Both underpin the traditional dichotomy familiar to our contemporary cultural consciousness, that is, that the Observer (and the various institutions he or she represents) actively monitors the populace, while we remain passively constructed according to their criteria. Magid draws attention to the ways in which equal treatment of persons within surveillance systems—specifically CCTV —effectively works against individualism, that unless a crime is committed, CCTV equalizes and obscures individuals, erasing them from memory. Furthermore, where a crime is committed and individuals are more selectively remembered, or when profiling or social sorting takes place, they are reconstructed from objective "memories"—highly discriminatory data flows—to fit particular narratives and agendas. Subjects are interpreted, outside of their agential control, and are cast as either criminals or victims. Any chance of positive, or indeed, self-determined subject-construction is likely to be fully negated. Jill Magid's responses to this negation have been presented in several of her artworks and writings, in which she employs a strategic form of dividual identity work to navigate surveillant spaces and practices. Magid's surveillant subject is deployed in order to explore and challenge the representation of bodies and identities by surveillance technologies through the appropriation of techniques normally allied with those technologies themselves. The surveillant subject is inherently dividual, in the Deleuzian sense, but essentially cyborgian, following Haraway, who proclaims that "we could hardly hope for more potent myths [than cyborgs] for resistance and recoupling" (1991, p. 154). Emerging from inchoate spaces rendered unexpectedly fertile in terms of their capacity to support viable identity work, Magid's subject resists by recoupling and restructuring itself from its disassembled components. In this way, it is newly able to reclaim agency and mitigates the negation of identity by circumventing the effects of equality and erasure.

ACTS OF APPROPRIATION: *LOBBY 7 (1999)* & *MONITORING DESIRE (2000)*

In her MSc thesis, submitted to the Massachusetts Institute of Technology (MIT), Magid stresses that "there is both a psychological and political value in reconstructing one's representation with the aid of technological systems" (2000, p. 5). Her attempt to reconstruct representation using CCTV challenges the way that identity, for which the body is an increasingly contingent element, is typically conveyed by surveillance cameras, and thus the way that the subjects under surveillance may reclaim agency. That reconstruction is necessary implies a consensus that identity is already deconstructed by surveillance environments, and that identity work must take place in order to re-establish the subject from those parts which have been undone. Magid writes:

> Within this system that begins with the surveillance camera and ends with the monitor, the body stands in the centre as both object and subject. Anybody stepping into this intermediary position goes through the same morphology – is subject to the same disorienting vision. In this system, the body becomes something unfamiliar, is represented by an alien perspective, and plays with a new set of spatial laws. (2000, p. 5)

Attempts to harness the creative potential of surveillance cameras in order to defamiliarize, alienate, or re-present the body are not new to contemporary or performance art. Both *Monitoring Desire* and a prototypical piece entitled *Lobby 7* draw heavily from the installation *Corps étranger* (1994) by British-Palestinian artist Mona Hatoum. In this earlier work, endoscopy and colonoscopy footage of the artist's body are projected onto the floor of a circular booth, while a "soundtrack recorded from the artist's internal organs" (Budgett 2001) provides audio to the viewing experience. *Corps étranger* depicts the artist's body made unfamiliar to her: fragmented and compartmentalized in line with the surveillant assemblage's perceptual impact upon bodies and subjects. Hatoum's piece makes for uncomfortable viewing, not solely for its "distressing pictorial effects" (Russell 1996, p. 1) but also due to the way it unsettles the traditional narrative that supports the self-as-known, therefore complicating the subject–object binary. This internal portraiture depicts aspects of the subject's body normally inaccessible to her; the new perspective offered by this self-surveillant technique reminds us just how much of ourselves remains unseen and thus

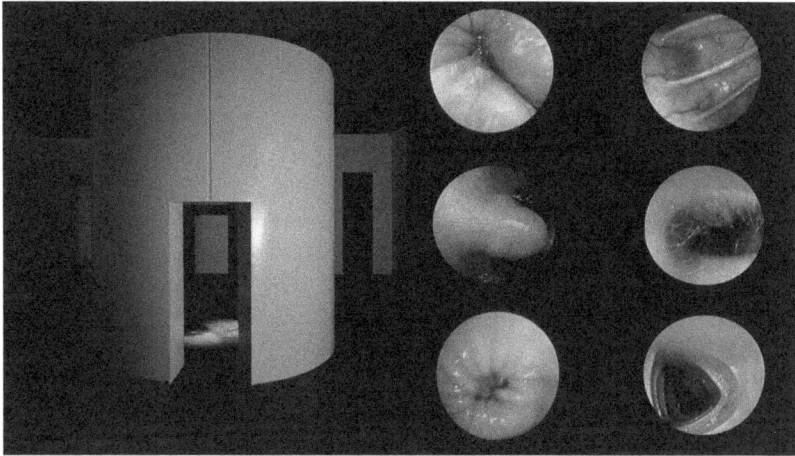

Fig. 2.1 (left) Mona Hatoum. *Corps étranger*. 1994. Video installation with cylindrical wooden structure, video projector, video player, amplifier and four speakers. 137 13/16 x 118 1/8 x 118 1/8 in. (350 x 300 x 300 cm). © Mona Hatoum. Photo © Philippe Migeat. Courtesy Centre Pompidou, Paris. (right) Mona Hatoum. *Corps étranger* (detail: film stills). 1994. Video installation with cylindrical wooden structure, video projector, video player, amplifier and four speakers. 137 13/16 x 118 1/8 x 118 1/8 in. (350 x 300 x 300 cm). © Mona Hatoum. Courtesy White Cube

unknowable, and further illustrates the dissolution of the body. The medically-surveilled subject is reduced to data flows which constitute situated knowledges that can only be truly parsed and made meaningful by authorised specialists. Hatoum is dislocated from her body, presented here as a fusion of viscera and information (Fig. 2.1).

Magid cites *Corps étranger* as one of her primary inspirations for *Monitoring Desire*, but where she most significantly diverges from Hatoum's approach is by combining the self-surveillant aesthetic with the methodologically significant appropriation of viewing techniques more normally regarded as unique to surveillance. *Monitoring Desire* was a performance piece enacted by Magid and fellow MIT student Orit Halpern at Harvard University's Science Centre, in which, "through a guerrilla act of appropriation" (2000, p. 2), recorded footage of the women was broadcast in real-time on the Centre's public informational monitor. Magid describes the captured footage as "produced by the camera on the shoe" that was worn by the performers in turn, and which "assimilated an abstracted view up the wearer's skirt with the surrounding architecture" (2000, p. 2). She continues:

> In the course of this performance, our bodies, as reconfigured through our surveillance apparatus, came to effect [sic] our subjectivities as they were presented in public space. Through the act of hijacking the informational monitor, we performed our power to publicly re-present ourselves back into the space in which we were occupying. (2000, p. 2)

The emphasis on the performance as a "hijacking" and a "guerrilla act" clearly points to Magid's appropriation of the surveillant space as an act of resistance, but this resistance is enacted in a perhaps unexpected manner. Rather than challenge the dissection that the surveillant assemblage performs upon subjects beneath its gaze, the challenge instead comes later, from within the system after having gained access. Magid appropriates the surveillant method by deconstructing herself using the same technologies, thereby mimicking the conditions of the system.

In an essay entitled "Theology of Mirrors", Magid insists that "if I can assimilate myself to a space, erase the boundary between space and my body, I will know what it is like to be imperceptible", claiming that "in Mimesis the environment is not an external feature but rather a definition of one's identity" (2002, p. 1). Jacques Lacan defines mimicry as "camouflage, in the strictly technical sense [...] not a question of harmonizing with the background but, against a mottled background, of becoming mottled—exactly like the technique of camouflage practised in human warfare" (1998, p. 99). Mimicry in these terms constitutes a mode of active resistance; the guerrilla act of appropriation signals an intervention by which the system may be surreptitiously breached, entered, and occupied. The combative form of mimicry encouraged by Lacan implies that erasure must first be embraced before it can be defied, and that from a concealed position the subject may then revisualise and reappropriate the surveillant space as a potent site from which to conduct identity work. In both *Monitoring Desire* and its prototype *Lobby 7*—a 1999 solo performance by Magid staged in the main lobby of MIT—the regular CCTV live feed was hijacked by the artist and her own footage broadcast in its place. Magid describes *Lobby 7* as an "exploration of my body and the surrounding architecture as seen through the natural openings of my clothes, via a lipstick surveillance camera that I held in my hand", a performance that lasted "one half hour—the time needed to capture every part of my body I that I could reach" (2005). The lipstick camera is small enough to be palmed, discreet enough to slip "inside" the subject and record its findings from within, mirroring the process of the artist herself, who has entered the surveillant assemblage and documents from a covert viewpoint.

The footage captured by the lipstick camera that comprises the video element of *Lobby 7* resembles the dislocated anatomical perspectives of *Corps étranger*, but in its exploration of the outer topography of the body the resulting imagery moves quickly beyond the biological or indeed the voyeuristic, instead taking on a surreal and faintly uncanny quality. The challenge of *Lobby 7* is the occupation and subsequent undermining of the subject-object relation, challenging not only to the conventions of the surveillant space, but also to the one experiencing the "disorienting vision" generated by the simultaneity of watching/being watched and the subversion of these long-established binaries that have conditioned self-perception. Once embedded within the system, having deconstructed herself in order to gain access and while simulating the system's conditions to remain camouflaged, the second act of Magid's performance was staged. While conducting a sweep of her body beneath her clothes, the artist stood directly in front of the monitor, observing the transmission of her exploration through this intermediary frame in real time and in full view of the public. That the performance lasted thirty minutes without interruption from MIT security—who no doubt were watching the same monitor—is evidence of Magid's capacity to dupe the system. In the footage, the dimensions of her body are so distorted by the camera's angle and proximity that one would assume that some time would have elapsed before passing spectators understood what they were seeing, despite the fact that the artist herself was conspicuously delivering the performance not twenty feet away from the compromised screen. Furthermore, in keeping with surveillant methods, the body itself was doubly abstracted from its contexts —as a body belonging to a subject, in the first instance, and as the series of images broadcast on a university CCTV transmission—which likely made it initially unclear that a breach in the system had occurred (Fig. 2.2).

Magid's second act of appropriation, this time at Harvard, employed a similar strategy of deconstructing and decontextualising bodies in order to gain access to a closed viewing system. In *Monitoring Desire*, the two female performers took turns to don the "surveillance shoe"—a high-heeled black leather sandal modified with a charge coupled device camera and wireless transmitter. While the wearer moved around the space

Fig. 2.2 Magid (background, partially obscured) conducts her self-exploration in *Lobby 7* (1999) while watching the image-capture in real time on the monitor

of the lower-level lobby, her partner remained on the first floor, watching the interrupting images transmitted from the shoe on a large CCTV monitor prominently situated for the benefit of the university community. Halfway through the performance, the wearer and the watcher exchanged the shoe, before retracing their movements in the opposite roles. While in *Lobby 7* the observer and the observed—and thus subject and object—are uniquely combined in the solo performer, in *Monitoring Desire* the introduction of a second agent, coupled with the location of the performers in relation to the monitor, limits the perspectives of the performers but also introduces new subjective positions that complicate the relationship between spectator and spectacle. Magid describes how:

> Our bodies as transmitted through our surveillance system become recon-
> figured in space. Because the wearer of the shoe is always downstairs, out of
> the monitor's view, her 'reconfigured' bodily construction is always invisible
> to her. In the Science Centre, the video image of my body and my physical
> presence are displaced from one another. The spectators have a choice: they

can either watch me unmediated by the surveillance camera downstairs or can view my virtual image in the monitor upstairs. The closest connection to seeing both positions at once is to look at my performance partner watching the monitor and imagine that she is a kind of stand-in for me. Or, later, that I am a stand in for her. (2000, p. 12)

In contrast to *Lobby 7*, where the act of appropriation, rather than the imagery itself, provides the central dynamic by which the piece's impact may be appreciated, the visual narrative of *Monitoring Desire* provides a second commentary on the nature of representation as it unfolds within the surveillant frame. Magid informs us that she is "specifically interested in the reconfiguration of women's bodies", noting that these bodies have "long been associated with concealment and issues of privacy" (2000, p. 15). The alignment of the female with the private, and the passive position in the viewing relationship, begins to reveal the ontological implications of these subversions in unseating the fundamentally dichotomous nature of viewing practices.

Binary categories—such as Observer/observed, authority/controlled, public/private, active/passive, desirer/desired, subject/object—are inherently supported by the surveillant space as it is traditionally conceived. To these pairs, male/female may be added in order to ameliorate our understanding of surveillant viewing practices and spaces by borrowing from feminist film studies' notion of the gaze (Mulvey 1999). In her writings, Magid repeatedly returns to these dialectical interplays which are inscribed upon, and ultimately reinforced, by the gendered bodies of surveillant subjects. If the male gaze dictates how we, as spectators, regard women when they are framed by cameras, then Magid responds by gendering the surveillant space as female; with women in the roles of both performer and observer, they regain control of how they are being represented. The initial destabilisation of the male/female dichotomy triggers the rapid breakdown of all other binary relations as the project exposes the overwhelming extent to which the male gaze is aligned with the authoritative Observer, the active subject desiring the one framed in the publicly surveillant space. As Magid concludes, "the reading of the performance as being potentially for two women, between two women [...] problematizes the spectator's possibly conventional notions of gender relations" (2000, p. 12), but it also problematizes the nature of the surveillant space as conforming to the accepted power-play symbolised by surveillance in general. With the elimination of the masculine viewpoint (the hijacking of

Fig. 2.3 The watcher observes the wearer in the neutralised space of *Monitoring Desire* (2000)

the monitor) and the substitution of the female perspective, the oppressive nature of the surveillant space is neutralised, and it becomes a space for the reconstruction of previously passive subjects and, as a result, a space of potential empowerment (Fig. 2.3).

In both *Monitoring Desire* and *Lobby 7*, Magid undermines the authority of surveillant technologies and practices by imitating their conditions in order to breach the spaces that they govern. In unsettling dichotomous relations that maintain this power, and by deconstructing her own identity to become a dividual subject composed of various data flows that enable her to embed herself within informational systems, the artist compromises the most fundamental dualism informing surveillance— the subject/object relation. The elusion of this dualism, upon which traditional constructions of identity have so long been dependent, invites a radical new form of dividual identity which is maintained through the resistant occupation of surveillant spaces. No longer constrained to defining their identities as one thing in opposition to another, a variety of new subjective positions are made available to the performers in the appropriated space—as Magid writes: "the system of our performance enables us […] to try on different identities and to share aspects of our identities with one another […] Our system is not one of constituting or validating a singular identity" (2000, p. 23).

In these performances, the subject is portrayed as reclaiming space in two fundamental ways. In the first instance, the guerrilla act of appropriation interpolates the subject into a closed system from which she would normally be excluded and erased. Secondly, once inside the system, the dividual subject draws her strength from imitating the conditions of the system which aims to contain her, and after voluntarily deconstructing herself into myriad flows of identifying information, proceeds to rebuild herself on her own terms. Using technological apparatus—cameras, monitors, screens—to express and enact this identity work renders the performers cyborgian, both in their resistant occupation of the technocratic control space and in the resulting compossibility of an identity woven from disparate strands of organic and technological information.

DIVERTING THE GAZE: *SURVEILLANCE SHOE/LEGOLAND (2000)*

The particular technique of image-capture used by Magid in *Monitoring Desire* and redeployed in her second self-surveillance project, *Legoland*, carries a powerful message regarding the reclamation of gendered identity in surveillant spaces. The upskirt camera shot is deliberately composed in order to further destabilise the voyeuristic power plays inscribed upon subjects by surveillance practices. Appropriating this angle, as well as the space itself, denotes an acerbic challenge to the conventions of the gaze as it draws attention to the distinctions drawn between power and empowerment (Fig. 2.4).

Significant research has been carried out on the upskirt shot, particularly in photography, by which means this voyeuristic image of the female subject is normally configured. Anne Allison has argued that such shots serve as a reinforcement of the gaze, and that this "position, particularly of males, as lookers, splits between one that is permitted and controlling and one that is illicit and immobilizing" (2000, p. 43). Allison draws her ideas from research into the upskirt or *appu-sukaato* image that is heavily fetishized in Japanese manga and anime—not only in adult and erotic publications, but also highly prevalent in animation aimed at children. The glimpse upskirt, she claims, enables the gaze to immobilise the subject, and this is supported by the medium itself—*appu-sukaato* images in manga

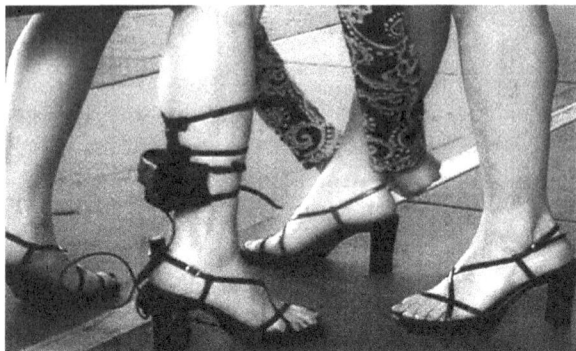

Fig. 2.4 The exchange of the surveillance shoe in *Monitoring Desire* (2000)

typically depict female characters held static and compliant in the frame of the panel, extradiegetically gesturing to the constraints of the male gaze. David Murakami Wood also acknowledges the trend of upskirt photography in Japan, claiming that such voyeurism, newly enabled by personal surveillance technologies (devices such as smartphones or webcams), has become "almost a cultural norm" (2005, p. 475). Noting the "climate of fear of such covert surveillance amongst women", he argues that "there is nothing one could regard as being positive about this particular form of 'people watching people'" (2005, p. 475).

It would appear that as institutional surveillance methods are rapidly democratised by the widespread use of camera-ready devices—which Zygmunt Bauman has labelled "personal panopticons" (2013, p. 59)—and previously stigmatised voyeuristic tendencies find new validation in popular culture, surveillant spaces generated by the immobilising gaze are becoming increasingly prevalent (but no less toxic) locations for the production of identity. In the upskirt image, female agency is erased even as her appearance is foregrounded—she occupies a paradoxical position of both amplified and silenced, emphasised and ignored. Magid's video projects with the surveillance shoe work to neutralise the voyeuristic bent of surveillant practices by re-empowering the subject of the gaze. By appropriating the static image of the upskirt shot and reinstating it in a kinetic medium, Magid inhibits the gaze and forces a new conception of the surveillant space by calling into question the nature of the bodies and subjects enframed within in.

Of the surveillance shoe as it functions in *Monitoring Desire*, Magid writes that "capturing the view up my skirt, one would expect a series of erotic images. Yet, much of the images' titillating effect is dampened by the disturbing sensations produced through their distortions" (2000, p. 16). As with the covert anatomical sweep of *Lobby 7*, the angle and proximity of the camera produces a sequence of discomfiting bodily images, dislocated as they are from the spectator's expectations of an erotically-posed upskirt shot. Magid notes that what unseats such expectations is the leg to which the camera is affixed, which "appears to be lame, making the body handicapped" (2000, p. 16). This tactic of distortion extends from the body as communicated by the camera to affect the status of the relationship between observer and observed in voyeuristic spectatorship; their previously stable roles are distorted, and in that moment of uncertainty the power is transferred to the actor in the frame. Moreover, as the artist so insightfully reminds us, "the camera, as we know, is not an objective eye but an eye with its own distorting practices" (2000, p. 16), which explicitly refers us back to the way that surveillance technologies work to erase the individual by breaking her down into data flows. Again, Magid is communing with the idea of the subject as intrinsically dividual, and she exerts control over the way she is represented by engaging in that process of deconstruction herself. Robert Knifton, in an essay situating Magid's work within the wider, real-world applications of surveillance, explains that "CCTV is generally highly open to interpretative narratives" (2010, p. 84). This is so because of the variety of flows that circulate throughout the frame: the restoration of a coherent narrative composed from a selective choice of available flows necessarily results in the final image being more or less distorted to suit the purpose for which it was initially generated. Magid mimics this method, and in doing so reveals the latent power in dividuality: that the space opened up by its interpretive potential provides a chance for the subject to reclaim her hold on that space, and thereby her own position within it, turning the situation to her subjective advantage.

Hille Koskela, writing about the sexualisation of surveillance, argues that in webcam culture, the shift from voyeurism to exhibitionism can be seen as "a form of resistance to the dominant gendered dynamics of monitoring", explaining that:

While the operator of a webcam cannot control who will see these images or how they will be interpreted, she or he is still able to control what, how and when these images are presented. Such revealing can be a form of political act which rejects the traditional understanding of objectification. (2012, p. 55)

David Bell, researching along similar lines, proposes an "erotics of resistance" by which the "radical potential of sexualized looking and being-looked-at" (2009, p. 210) enables the re-embodiment of surveillance. Framed by such statements, Magid's surveillance shoe becomes an instrument of resistance, a way to reinstate agency in the surveillant space and simultaneously upset the dialectics of power that normally underpin and maintain the structuring of such spaces. Koskela (2004) encourages subjective emancipation via the desexualisation of surveillance; Bell (2009) claims that the same resistant ends can be achieved through the re-emphasising of surveillance's inherent and creative support for sexual expression. Magid, channelling the cyborgian ethos of "holding incompatible things together because both or all are necessary or true" (1991, p. 149), manages to do both, and furthermore invites new readings of the relationship between subjectivity and spatiality enacted beneath the surveillant gaze.

The space itself, as the stage for the performance taking place, informs the changing nature of the surveillant subject. Where surveillance often abstracts the body from its surroundings in order to more effectively reduce it to decontextualized data flows, Magid reemphasises the relationship between self and environment by confusing the boundaries between, and characteristics of, the two (Fig. 2.5).

In *Legoland*, a 6-min video recorded as Magid walked through an unnamed city at night, the surveillance shoe captures the synchronised distortion of "the interior space beneath my skirt and the exterior space around the skirt's circumference" (2000, p. 30). Magid describes how:

The transgressive gaze up the skirt is difficult at first to get beyond. The space outside of the skirt becomes active to the viewer only after the voyeuristic novelty of this perspective passes. Because the image's distortion and the body's appearance of being crippled, it does not take long for the image to lose its overtly sexual quality. Emphasis gets passed to the strangeness of the space surrounding the body. The space appears to be tied to the body, even as a victim of it [...] One leg is always bound within the frame. This is the leg to which the camera is attached. Because of its placement, the camera seems to anchor my body in place. While this appendage appears as fixed,

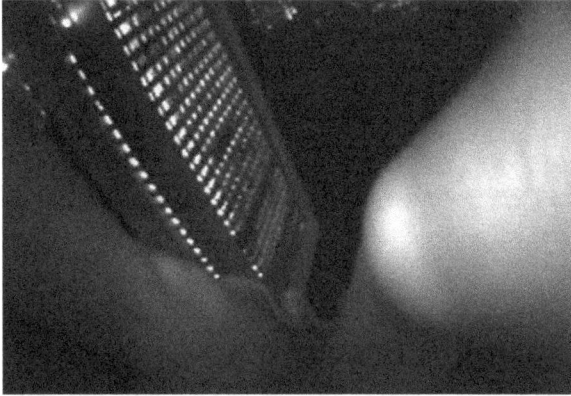

Fig. 2.5 Upskirt shot of the subject blending with the surrounding architecture in *Legoland* (2000)

everything else – including my other leg – is in motion. There is a strong inversion in that while the one leg appears to be stable like architecture, the actual architecture becomes mobile. It is as if the space in which I move is on a scroll, and with the kick of my "free" leg, I am able to unroll it. (2000, pp. 30–31)

Here, not only is the subject foregrounded within her environment, but the boundaries between body and architecture become less clear. The body acts as architecture, while the architecture is "activated and warped" (2000, p. 32) by the lens and thereby becomes far more flexible, almost organic in its elasticity. The surveillant space, tied to the leg of the subject, is directed by her, forced to follow, and the gaze is not only inhibited, but diverted, as the erotic sensation produced by the images quickly subsides. What remains is a performative interaction between a subject and the space in which she is newly reinstated, on her own terms. The data flows of the dividual subject circulate throughout the body and its surroundings in such a manner that mimicry and camouflage are exchanged for empowerment and presence.

Cyborgs take great pleasure in transgressing boundaries, embracing erasure in order to become "ether, quintessence" (Haraway 1991, p. 153), to dissolve into data flows and disappear into the bowels of the system, from which point they may initiate the process of dividual reconstruction that ultimately enables them to reinstate their claim to agency. Magid's projects, in unambiguous response to cyborg theory, repurpose surveillant practices to present the deconstructed, compartmentalised subject as liberated in her dividuation. This subject has traditionally been the victim of the equalising and erasing procedures encouraged by surveillant viewing, forcibly taken apart and taxonomised and finally re-presented via the male gaze as an object or indeed collection of depersonalised objects to be watched on-screen. Magid reaches out to the disempowered subject and offers her a way to reconstruct that representation, and in doing so to neutralises the gaze that pins her within the frame. This empowerment is achieved through the strategy of appropriating the very technologies and techniques that have been used to oppress, contain and negate subjects of the spectacle, for whom ubiquitous surveillance has significantly reinforced the limitation on spaces in which to enact viable identity work.

CONCLUSION

In *Discipline and Punish*, Foucault outlines the effect that panoptic control exerts over the subject—a force which Gary Marx later termed "soul training" (2009, p. 378). Marx meditates upon the efficacy of artistic representation for the social scientific scholarship of surveillance, posing several incisive questions:

> Artistic creations can significantly inform us about surveillance and society [...] Artistic statements, unlike scientific statements, do not have to be defended verbally. But the social scientist can ask about their social antecedents and impacts. Do they move the individual? Do they convey the experience of being watched or of being a watcher? Do they create indignation or a desire for the product? Do they make the invisible visible?" (2009, pp. 389–390)

Magid's work goes some way to providing answers to Marx's queries, and in doing so, treats the Foucauldian concept of panoptic soul-training by offering radical new perspectives on the ways in which bodies, behaviours and subjectivities are affected by surveillant practices of domination, self-governance, and control.

Magid sees her own art practice as a series of "social engagements that propose new relations, and thus new meanings, within existing social and public systems of authority. This includes their subversion" (2007). Her work bridges the gap between the art world and technoscientific disciplines, between academia and popular culture, and such subversive, unique and yet relevant approaches to surveillant systems continue to be developed in her later pieces. Scholars from a range of disciplinary orientations have, following Foucault, focused on the shifts in the modification and self-governance of behaviour and identity enforced by panoptic control, but the question has prevailed as to whether, within the close dialectical confines of the surveillance society, any opportunities arise to reconfigure surveillant spaces as spaces of empowerment. Magid's work responds to this concern in the affirmative.

Haraway maintains that her cyborg, above all, "can suggest a way out of the maze of dualisms in which we have explained our bodies and our tools to ourselves" (1991, p. 181), which in the context of Magid's work offers a way to revisualise the dichotomies of power and control underpinning post-Orwellian, post-Foucauldian understanding of surveillant spaces and the institutions and tools that construct them. Moreover, the cyborg destabilises the grand Western metanarratives of self, through its dividuality and its close kinship with both organism and machine. Its subversive potential for appropriating technologies of domination draws attention to the fact that, historically, "myth and tool mutually constitute each other" (Haraway 1991, p. 164). Magid sees new subjective potential in the harnessing of technologies previously used to monitor, separate, and sort individuals. She writes that:

> To be human today is to be totally intertwined with technology, specifically with the technology of image-capture. I am exploring this captured space under the eye of surveillance as a platform for the formation of new identities. In the performance, we as performers realize the potential of our appropriated surveillance technology to function as a vehicle for our empowerment. We chose to step into the line of this appropriated vision in order to frame ourselves differently. (2000, p. 21).

What might we see, if we manage to see ourselves differently via surveillance? Ontologically speaking, perhaps what surveillant spaces and practices allow us to re-present to ourselves is the dividual nature of our identities that was always already emerging in resistance to the negation of the contemporary subject.

REFERENCES

Allison, A. (2000). *Permitted and prohibited desires: Mothers, comics, and censorship in Japan.* Berkeley, CA: University of California Press.

Ball, K., & Haggerty, K. D. (2005). Editorial: Doing surveillance studies. *Surveillance and Society, 3*(2/3). Retrieved December 23, 2015 from http://library.queensu.ca/ojs/index.php/surveillance-and-society/article/view/3496/3450.

Bauman, Z. (2006). *Liquid modernity* (6th ed.). Cambridge: Polity Press.

Bauman, Z., & Lyon, D. (2013). *Liquid surveillance.* Cambridge: Polity Press.

Bell, D. (2009). Surveillance is sexy. *Surveillance and Society, 6*(3). Retrieved December 23, 2015 from http://library.queensu.ca/ojs/index.php/surveillance-and-society/article/view/3281/3244.

Bigo, D. (2008). Globalized (in)security: The field and the ban-opticon. In D. Bigo & A. Tsoukala (Eds.), *Terror, insecurity and liberty: Illeberal practices of liberal regimes after 9/11* (pp. 10–48). Oxon: Routledge.

Brighenti, A. M. (2010). Artveillance: At the crossroad of art and surveillance. *Surveillance and Society, 7*(2). Retrived December 23, 2015 from http://library.queensu.ca/ojs/index.php/surveillance-and-society/article/view/artveillance/artveil.

Budgett, G. (2001). *Introduction to spatial practices.* Retrieved January 11, 2016 from http://www.arts.ucsb.edu/faculty/budgett/classes/art12/hatoum.html.

Calle, S. (1980). *Suite vénitienne.* Set of 81 elements, made in this form from 1996 onwards: 55 b/w photographs, 23 texts, 3 maps. Edition of 3.

Calle, S. (1981). *The Shadow.* French version: diptych, texts and colour and b/w photographs, each element framed. English version: set of one text, one colour photograph, 29 b/w photographs partly assembled in groups, 11 texts. Edition of 3+3.

Deleuze, G. (1992). Postscript on the societies of control. *October, 59,* 3–7.

Deleuze, G., & Guattari, F. (1987). *A thousand plateaus.* Minneapolis: University of Minnesota Press.

Foucault, M. (1977). *Discipline and punish: The birth of the prison.* New York: Vintage.

Gandy, O. (1993). *The panoptic sort: A political economy of personal information.* Boulder: Westview.

Giddens, A. (1990). *The consequences of modernity.* Stanford: Stanford University Press.

Haggerty, K. D., & Ericson, R. V. (2000). The surveillant assemblage. *British Journal of Sociology, 51*(4), 605–622.

Haraway, D. J. (1991). *Simians, cyborgs, and women: The reinvention of nature.* London: Free Association Books.

Hatoum, M. (1994). *Corps étranger* 1 cylindrical structure, 1 video projecter, 4 loud-speakers, 1 video tape, PAL, colour, stereo sound, 30'.

Knifton, R. (2010). You'll never walk alone: CCTV in two liverpool art projects. In O. Remes & P. Skelton (Eds.), *Conspiracy dwellings: Surveillance in contemporary art.* Newcastle-upon-Tyne: Cambridge Scholars Publishing.

Koskela, H. (2004). Webcams, TV shows and mobile phones: Empowering exhibitionism. *Surveillance and Society, 2*(2/3). Retrieved December 23, 2015 from http://library.queensu.ca/ojs/index.php/surveillance-and-society/article/view/3374/3337.

Koskela, H. (2012). You shouldn't wear that body: The problematic of surveillance and gender. In K. Ball, K. D. Haggerty, & D. Lyon (Eds.), *Routledge handbook of surveillance studies.* Oxon: Routledge.

Lacan, J. (1998). *Book 11 of séminaire de jacques lacan* (2nd ed.). New York: W. W. Norton and Company.

Lyon, D. (Ed.). (2003). *Surveillance as social sorting: Privacy, risk, and digital discrimination.* London: Routledge.

Magid, J. S. (2000). *Monitoring desire.* MSc., Massachusetts Institute of Technology. Retrieved August 22, 2011, from http://dspace.mit.edu/bitstream/handle/1721.1/76084/47864809-MIT.pdf?sequence=2.

Magid, J. S. (2000). *Legoland.* B/W footage produced by Surveillance Shoe. 6 min.

Magid, J. S. (2002). Theology of mirrors. Retrieved March 10, 2010 from http://www.jillmagid.net.

Magid, J. S. (2007). *Seduction.* Retrieved March 10, 2010 from http://www.jillmagid.net.

Marx, G. T. (2009). Soul train: The new surveillance in popular music. In I. Kerr, V. Steeves, & C. Lucock (Eds.), *Lessons from the identity trail: Anonymity, privacy and identity in a networked society.* Oxford: Oxford University Press.

Mulvey, L. (1999). Visual pleasure and narrative cinema. In L. Braudy & M. Cohen (Eds.), *Film theory and criticism: Introductory readings.* New York: Oxford University Press.

Murakami Wood, D. (2005). Editorial: People watching people. *Surveillance and Society, 2*(4). Retrieved December 23, 2015 from http://library.queensu.ca/ojs/index.php/surveillance-and-society/article/view/3358/3321.

Pridmore, J., & Zwick, D. (2011). Editorial: Marketing and the rise of commercial consumer surveillance. *Surveillance and society, 8*(3). Retrieved December 23, 2015 from http://library.queensu.ca/ojs/index.php/surveillance-and-society/article/view/4163/4165.

Rosen, D., & Santesso, A. (2013). *The watchman in pieces: Surveillance, literature, and liberal personhood.* New Haven: Yale University Press.

Russell, S. (1996). Corps étranger. *Zing Magazine,* [online] 1 February. Retrieved January 11, 2016 from http://www.zingmagazine.com/zing3/reviews/020_body.html.

AUTHOR BIOGRAPHY

Amy Christmas is an Assistant Professor at Qatar University in the Department of English Literature and Linguistics. Her research explores the human aspect of technology in science fiction and culture and her work on Jill Magid has been presented at Liverpool University and the University of Tartu.

CHAPTER 3

Camera Performed: Visualising the Behaviours of Technology in Digital Performance

Jaclyn Meloche

Recording thus had the strange effect of capturing the potentially unpredictable behaviour of the artist-performer and suddenly rendering it into an object outside of time and space. (Salter 2010, p. 116)

In *Entangled: Technology and the Transformation of Performance*, media artist and Professor Christopher Salter speaks to the transformative and anthropomorphized function of the camera arguing that its frame is no longer an aesthetic device that historically was introduced to outline the contours of a stage, but that the mechanical lens in fact co-opts a performative role in the reproduction and dissemination of meaning in contemporary visual culture. In this chapter, I draw from Salter's research to outline a theoretical framework for understanding the ways in which performance, performativity and digital performance collectively complicate the inherent function of the webcam and the mechanical lens in digital media art. With reference to poststructural theory and contemporary art

J. Meloche (✉)
Art Gallery of Windsor, Windsor, Canada
e-mail: jacmeloche@bell.net

S. Flynn and A. Mackay (eds.), *Spaces of Surveillance*,
DOI 10.1007/978-3-319-49085-4_3

historical discourses pertaining to media art and lens culture, particularly in the writings of Christopher Salter and media choreographer Johannes Birringer, I reconsider how the digital lens performs a *détournement* through the power of the gaze. Moreover, in response to the central question in this collection—how has surveillance permeated contemporary society?—I look at the ways in which the webcam acts as a transformative agent that mediates reality in new media art.

Exemplified in *Glenlandia*, a digital media work, by British artist Susan Collins, the webcam becomes an embodied instrument that reproduces an identity through surveillance, mapping and digital crossover visualization. Collins, who uses networked tools to problematize the mimetic behaviours of representation, appropriates digital technology to visually and materially blur real-time and virtual-time. Through a metaphorical and social deconstruction of the webcam in contemporary art practice, I challenge the notion of digital representation by arguing that networked practices are inherently dialogical and performative.

Performance, according to its definition in the *Encyclopedia of Aesthetics*, is defined as "the execution of a range of 'artistic making and doing'" (Alperson 1998, pp. 464–466). Although too simplistic a definition for such a historically loaded medium, what is of interest in this description, as well as in my research, is the idea of execution and how nonhuman matter becomes performative. In his analysis of performance and performativity, Salter focuses on the performative nature of process. In his text, he maps a dialogical lineage for understanding the *entangled* relationships between performance and technology in contemporary art historical and scientific discourses. By asking the question *how*, he challenges the various meanings of performance, performativity and digital representation in order to reconsider their functions and aesthetics as fundamentally interconnected matter. Consequently, the premise of his text asks the following two questions: how does performance merge real-time and digital-time? And how can a technical and inanimate object embody performativity?

The term performance, he explains, is deeply rooted in the dematerialization of the object; "the label *performance* was a strategy used to describe actions, happenings, and time-based events emerging out of the visual arts during the 1950s through the 1980s" (Salter 2010, pp. xxiii–xxiv). In reference to an experience, this framework suggests that art, more specifically performance art, has moved away from what art historian and critic Michael Fried once argued is the essence of fine art: *objecthood*,[1] and instead towards an experience in which there is a triangular exchange between the artist, the subject and real-time. Although this definition of performance suggests an inherent need for action and movement within a work of art, it is arguably its adaptation to change that remains at the core of performance as a creative medium.

The intertwined relationship between the medium and the experiential is an innate component of performance. By reenacting a sensorial experience of embodiment, the performer enters into an active engagement in which they themselves become an actant in an exchange. In this instance, the performer begins to embody performativity through the naturalization of the body in time and space. In other words, the performative subject is one whose identity is not fixed, but rather dialogically reproduced through a co-constitutive relationship.

In her analysis of performance and performativity, feminist cultural theorist Judith Butler explains that an identity is not neutral but rather constructed from a series of acts; "… performativity is not a singular act, but a repetition and a ritual, which achieves its effects through its naturalization in the context of a body, understood, in part, as a culturally sustained temporal duration" (Butler 1990, p. xv). Rooted in the fields of gender studies and poststructural feminism, Butler insists that an identity is not innately inherent to gender, more specifically woman, but that an identity is the result of a sociocultural performative exchange (Kavka 2001, p. xiv). According to this definition, performance therefore represents a series of actions and exchanges that occur through moments of performativity.

In keeping with Butler, cultural theorist and critic Mieke Bal describes a performative act as one that is endlessly repeated and ritualistic, and that through its perpetuation, the identity of the performer, or subject, is essentially re-constructed. Furthermore, she explains that "[p]erformativity, on the other hand, is allegedly the unique occurrence of an act in the here-and-now" (Bal 2001, p. 110). Framed in the context of modification, performativity becomes the result of an exchange that over time and through repetition re-materializes what already exists. In digital performance, performativity is best understood through the machinic entanglement of technology and human interaction. The mechanical tools used to represent and recreate meaning in digital performance thus become performative in their perpetual crisscrossing between mediums, technologies and platforms.

Lens Culture in a Technosocio-scientific Context

The relationship between the machine and performativity in contemporary art theory is commonly characterized as an anthropomorphized, humanized, metaphorical and technosocio-scientific exchange. In his research, cultural anthropologist and Professor David Howes argues that an object is not neutral, but rather a social and cultural construct. In "Charting the Sensorial Revolution," Howes asserts that objects are "bundles of sensory properties and interconnected experiences that activate the human senses in complex and culturally varied ways" (Howes 2006, p. 115). In other words, that the machine is not singular, but rather a dialogical force in the creation of an experience.

To further exemplify this poststructural reading of technology in digital media art, Salter borrows from Félix Guattari's definition of the machine to further emphasize the notion that the machine is, in fact, a productive assemblage of "*components* that are co-constitutive with each other" (Salter 2010, p. xxxiii). In theory and practice, an assemblage represents an active form of reconstruction that is likened to a repurposing of materiality. In her text *Vibrant Matter: A Political Ecology of Things*, political theorist Jane Bennett describes an assemblage as an "ad hoc grouping of diverse elements, of vibrant materials of all sorts. Assemblages are living, throbbing confederations that are able to function despite the persistent presence of energies that confound them from within" (Bennett 2010, pp. 23–24). Digital media art and, more specifically, digital performance therefore parallel the framework of an assemblage. By representing a reproduction of real-time, the live matter that is dialogically interconnected through cyberspace offers a modified interpretation of liveness.

In Salter's material historicity of digital performance, he further explains that the machine is relevant to the medium through its ability to adapt to change; "[d]espite the different connotations of performance in the technosocio-scientific context, the move towards agencies, collectives, and networks articulates a common thread: that humans, things, and matter are not fixed but always in a process of change and becoming" (Salter 2010, p. xxx). Salter's argument is thus that through the interconnectivity of networks and perpetual interaction, the meaning of matter is reconsidered and reproduced. In keeping with Butler and Bal's analysis that an identity is not innately fixed, he argues that the function and presence of performativity in this context exist as fluid and co-constitutive, and therefore subject to change and reinterpretation.

Christopher Salter's investigation of performance vis-à-vis performativity suggests that although the two theoretically work hand-in-hand, it is the performative nature of the machine that informs the ways in which he contextualizes performance in contemporary media art. In his research on modernism and the rising powers of technology, Swiss architectural historian Sigfried Giedion premises the notion that the machine is a metaphorical extension of the human body; that the machine, as a dialogical material object, adopts human behaviours when active and live. In other words, that "the hailing characteristic of the machine, above all else, as the phenomena of movement itself [is] an extension of the human body and an articulated, performative substitute" (Giedion 1969, pp. 46–50). Aesthetically and biologically technological, the machine, however, becomes an animate matter. Able to interact and perform human-like behaviours, the machine in digital performance thus becomes a stand-in for the body.

The cross-cultural function of the machine in digital media art is also the subject of Johannes Birringer's writings in which he seeks to establish an understanding for the new relationship between the body and technology in digital performance. In essence, Birringer sets out to define digital performance in relation to the performative nature of what he terms the instruments of new media technologies writing that "digital performance [is] a term [he wants to use] for performances that depend on the use of digital interfaces" (Birringer 2008, p. xii). In support of a new way of looking, he is working to understand the entangled relationships between digital aesthetics, performance experimentation and technological development. Consequently, his is an interpretation of the interconnected relationships between humans and technology, and aesthetics and science that informs the role of technology in current discourses on digital contemporary art (Birringer 2000).

Moreover, Birringer contextualizes a working definition for digital performance by identifying its dependency on the interchange of digital interfaces. More specifically, that the interface, as a visual platform for performativity, provides a space for the aesthetic and conceptual reconstruction of real-time that is inherent in digital practices. To complement this concept, Birringer further notes that the evidence of performativity, through technology and the interface, is represented through interacting behaviours; "[t]he convergence between performance and technology reflects back on the nature of movement and behaviour, and particularly on the nature of 'body' and our understanding of its objecthood or identity;

its organization and augmentation; its physical-sensory relationship to space and the world; its immediate, phenomenological embodiedness but also the inseparability of its embodiment from the technical" (Birringer 2008, p. xxii).

In response to Birringer, I ask: is the execution of performance in new media digitization the result of a dialogical exchange on the interface, or does the interface provide what can be considered a platform for the distribution of digital aesthetics through performance?

The notion that the interface is a platform for the dissemination of meaning is contingent on the dialogical relationship between the webcam, the Internet and the computer. Best described as a miniature digital video camera, the webcam, when connected to a computer, streams live digital images of its subject through the Internet and onto a webpage. By taking photographs at set intervals, that can range from 5 s to 30 min apart, the webcam records and distributes pixilated representations of its subject in real-time as seen through the lens of the camera.

SUSAN COLLINS AND AN ACT OF SURVEILLANCE

Susan Collins, a digital and media artist based in the United Kingdom, appropriates the technology of the webcam to create digital recordings and photographic documentations of site-specific locations. Through detailed digital programming and designed systems of pixilation, the artist captures an image of real-time through a systematic construction of squares. In a detail from the series *Glenlandia* (2005–2007) (Fig. 3.1), the artist's mechanical construction of pixal-generated imagery is exemplified in a visual deconstruction of real-time. To create this piece, Collins installed a webcam at the Fisheries Research Services (FRS) in Faskally, and over a period of two years, recorded the scenes of Loch Faskally in Perthshire, Scotland.

On her website, the artist describes the recording process as follows; "[The] webcam harvested images pixel by pixel, second by second and day-by-day over the course of two years. Each image was collected from top to bottom and left to right in horizontal bands continuously, marking visible fluctuations in light and movement throughout the day and archived at two-hour intervals" (Collins 2005–2007). Constructed from layers of coloured data, the images over time become modified and transform into suggestive documents that depict the behavioural effects on place.

Fig. 3.1 Susan Collins, *Glenlandia*, 2nd June 2006. Digital Image from Live Transmission

According to the literature available on the subject of the webcam, new media historian Sheila Murphy notes that the webcam is not only a tool used to monitor, record and transfer digital images of real-time, but has also become a cultural phenomenon that has induced a sense of fear and cultural ubiquity (Murphy 2000). In "Lurking and Looking: Webcams and the Construction of Cybervisuality" Murphy sets out to contextualize the webcam as a performative signifier for the construction of cultural engagement through the Internet. Drawing from theories on surveillance and what American art historian and critic Rosalind Krauss names the single most important statement yet made about photography (Krauss 1999), Walter Benjamin's essay "The Work of Art in the Age of Mechanical Reproduction" which questions the social function of art—particularly reproducible art, Murphy maps a narrative that complicates the webcam as a purely mechanical device, introducing instead the notion that its inherent function is performative and instrumental in the recreation of meaning; "[t]he 1990's has been a decade in which communications technologies and moving image technologies have converged in multiple ways, producing new cultural extensions" (and fragmentations) of the human body and of subjectivity" (Murphy, p. 173). Thus, by making the technology of the device transparent, Murphy strips the camera down into a metaphorical part of the human body.

The machine as body analogy is described in contemporary art theory as a mechanical apparatus that acts as a metaphorical, and all seeing, prosthetic eye. For example, Canadian art historian Kirsty Robertson likens the webcam to a technological eye, a pair of glasses and lens that ultimately scan and survey real-time in real-place. Although she notes that surveillance quite literally means to inspect, examine and scrutinize, it is also indicative of an animate way of looking (Robertson 2010). Arguing that the camera represents a prosthetic eye that watches and records its subject, she convincingly explains that the lens of the camera becomes a performative device that watches, re-enacts and repurposes what it sees.

The performative nature of the instrument is further exemplified in the modification of meaning and identity that is made possible through the gaze. In feminist discourses, the gaze is understood as a gendered power struggle between the male gaze and the female subject. By way of watching, the male gaze subjugates the female's power and therefore imposes, through looking, a social and sexual hierarchy. The exchange of the gaze in surveillance inevitably insinuates a social and political hierarchy, but it also paves the way for a reactionary methodology. In the spirit of a *détournement*, a reactionary exchange that casts a new purpose onto a subject, Robertson argues that the surveying "eye", in theory, performs a similar reconstructive function: through the act of looking and recording, the mechanical lens alters the original nature and identity of its subject.

In keeping with Robertson, art historian Sarah E.K. Smith interprets surveillance as a performative act in which the exchange of the gaze ultimately shifts a subject's identity through perpetual looking. In "Captured and Controlled: Critiquing Surveillance through the Camera" Smith contextualizes the role of the gaze arguing that, "… because artists appropriate rather than return the gaze, [video] art is able to offer new insights to discussions of surveillance. In other words, that surveillance reproduces our ways of looking by distancing the viewer and changing how we look" (Smith 2010, p. 60). Similar in theory to Judith Butler's concept of performativity, in which the repetition of an act ultimately reconstructs an identity, the surveillant gaze too becomes performative in that the non-stop act of looking modifies its "real-time" identity.

Born from the notion that looking is a form of power, surveillance is essentially a "system that monitors the actions of its populace" *We Know You Are Watching: Surveillance Camera Players* 2006, p. 21. To add to this definition, sociologist and Professor David Lyon explains that although surveillance is a form of collecting personal data, it is more so about the act of exchanging gazes when watching or being watched (Lyon 2001). Moreover, he boldly suggests that surveillance is not a capitalist ploy to monitor and induce fear, but rather that today surveillance is an inherent component of today's computer generation. Moreover, that "the most important means of surveillance reside in computer power, which allows collected data to be stored, matched, retrieved, processed, marked and circulated" (Lyon 2001, p. 2).

Over the last twenty years, Lyon has interpreted surveillance in the contexts of behavioural psychology, poststructuralism and the machine aesthetic arguing that its performative gaze results in the transformation of a subject or body when it is viewed through a mechanical lens. In his own words, "[w]hen computer-power and sophisticated statistical technique come together, all sorts of profiles of persons and populations can be built" (Lyon 1994, p. 84). An expert on discourses of surveillance, he further lends his research to studies of the gaze and its influence on human interactions and social behaviours. For example, he argues that an identity is a dialogical construction that is contingent on others' perceptions; "[o]ur identity is understood by others—and by inanimate machines—more from our data-image than from our personal communication" (Lyon 1994, p. 19). In keeping with Robertson and Smith, Lyon's framework for surveillance argues that, as a form of visual culture, the act of watching is not synonymous with viewing real-time. Instead, surveillance, as a gathering system, stores information and uses technology to modify its digital reinterpretations.

The webcam, the digital instrument of interest in this inquiry, is inoperable without its sister platform, the World Wide Web. Dependent on the Internet for its transmission of information and visualization, the webcam's mechanical function is meant to transfer recorded imagery through an online server that can then be viewed, in theory, on any computer screen. According to media artist Andrea Zapp, the Internet represents two major networked platforms: one that is communicative and the other that is participatory. In "A Fracture in Reality: Networked Narratives as Imaginary Fields of Action," Zapp explains that the "net is a comparatively unique cosmos of invented identities, partakers, and accomplices in joint forces, hidden in the endless labyrinth of home pages, chatrooms, and communities" (Zapp 2004, p. 181). Metaphorically speaking, the Internet is therefore a virtual space in which the fluid interchange between time, space and place continuously blur one's perception of reality.

PRACTICING DIGITAL VISUALISATION

Exemplified in her practice, Collins appropriates the Internet as a tool for the dissemination of digital visualization and meaning. Through the process of online streaming, she creates a path for the transference of imagery between real-time and virtual-time. The Internet, in the context of Collins' digital recordings, is not only used to connect and reconstitute virtual platforms, rather it also becomes a tool used to freely communicate and view.

In a description of *Glenlandia* (Fig. 3.2), the artist specifically notes that the web serves to enable a sense of freedom for viewing; "[*g*]*lenlandia* was developed as a distributable artwork—a full screen landscape which when the work was transmitting live could be displayed full screen installed in a gallery with the work updating in real-time. It was also possible for viewers to download the relevant display software from this page for a full screen view on their home computers" (Susan Collins 2005–2007). The ability to watch a digital recording of place in real-time in the comfort of one's home is a perfect parallel for the argument that the Internet represents freedom. Because of the web, a person no longer has to leave their house to "travel." Likened to an armchair traveller who travels through photographs, a web traveller can view anything, anytime and anywhere.

To complement the idea of virtual travel is the act of mapping place in cyberspace. The digital map is arguably an inherent component of the Internet that ultimately acts to abstract reality by blurring its aesthetic and locative characteristics. In cyberspace, every Internet website has an unrecognizable address that positions its whereabouts. However, the abstract and unclear coding of a web address essentially "unmaps" a place rather than situates it.

In "Online Performance: "Live" from Cyberspace," actor and scholar Steve Dixon negotiates the relationship between real-space and cyberspace arguing that an online address is both arbitrary, yet site-specific; "[a] URL (Uniform Resource Locator) may be a Web address whose arbitrarily lexicon gives no clue to the geographical positioning of its real-place server, but it can also be seen as a type of map coordinate, defining a specific location and meeting place in the geographical contours of the network"

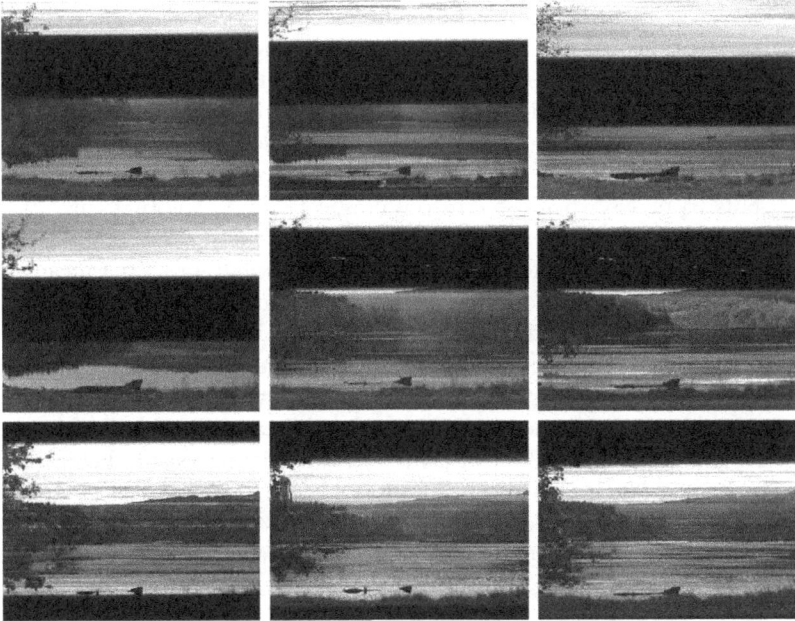

Fig. 3.2 Susan Collins, *Glenlandia*, (2005–2007). 9 Digital Images from May 2006

(Dixon 2007, p. 463). In keeping with Zapp, who argues that a web address signifies every place in cyberspace, Dixon explains that geographical boundaries become blurred through an abstract locative digitization of online mapping. Further perpetuating the argument that place is performative and subsequently that virtual place is not fixed, but rather fluid and dialogical, he suggests that the abstract characteristics of the digital map blur the boundaries between real-place and virtual-place.

In keeping with Dixon's argument, landscape architect James Corner further notes that mapping introduces a new and fresh way for looking at the world, writing that, "… the function of mapping is less to mirror reality than to engender the re-shaping of the worlds in which people live" (Corner 1999, p. 213). Therefore, subject to constant relational change, the act of mapping place in cyberspace transforms the interface into a performative stage that continuously shifts in meaning and aesthetics through a dialogical exchange between real-place and virtual-place.

Similar to Corner's analysis of online mapping, *Glenlandia*, as a digital map, also proposes a new way for viewing, mapping and experiencing place. By consciously complicating the notion that the camera is a mirror, Collins, in an interview with Carlo Zanni admits that her online representations of place are fluid and without geographical boundaries; "...rather than simply challenging notions of 'real' or 'virtual'—is the possibility the online world presents to be elsewhere—so in some respects I am exploring (my own) fantasies relative to that (in *Fenlandia* as with much of my other work) whilst simultaneously attempting to expose the reality of it (the means of production) as embedded in the work itself" (Zanni 2006).

Entangled in a blurred reproduction of fact and fiction, why then, is the artist so specific in her identification of the webcam's whereabouts? When contextualizing *Glenlandia* on her website, the artist describes in detail the specific points of reference for each recorded location writing that "Loch Faskally, a quintessentially Scottish man made lake, is located behind the hydro dam at Pitlochry which was built in 1947–1950 as part of the North of Scotland Hydro-Electric Board's Tummel/Garry Power Scheme" (Collins 2005–2007). In other words, I ask, if these details were not available, would it be possible to locate this site based on the digital image or the online web address? Probably not. Therefore, the act of mapping does not project a concrete outline of real place, but rather suggests an abstract experience of the source.

In their interpretation of the map as a spatial experience, philosophers Gilles Deleuze and Félix Guattari argue that the effectiveness of mapping lies in its ability to represent one's interaction with place. Moreover, that a map literally and materially offers a platform for performance; "[W]hat distinguishes the map from the tracing is that it is entirely oriented toward an experimentation in contact with the real. The map does not reproduce an unconscious closed in upon itself; it constructs the unconscious. It fosters connections between fields, the removal of blockages on bodies without organs, the maximum opening of bodies without organs onto a plane of consistency...The map has to do with the performance, whereas the tracing always involves an alleged 'competence'" (Deleuze and Guattari 1987, pp. 12–13).

To complement Deleuze and Guattari's concept that mapping represents a visual experience of place, Tyler Mitchell claims that the interactive nature of web maps make them more accurate than a painted or photographed map. Because virtual images are continuously updated, Mitchell insists that its formal and spatial development and organization of land are;

therefore, more likely to mimic the "true" nature of a place (Mitchell 2005, pp. 1–3). If this is true, then how does one accurately document performance, albeit digital or live?

Susan Collins, in her practice, negotiates the issue of visual documentation through photography. In an installation view from the exhibition *Outlook Express(ed)*, hosted by Oakville Galleries in Oakville, Ontario in 2007 (Fig. 3.3), *Glenlandia* is represented in a series of digital photographs. The appropriation of photography to record real-time, according to Birringer, is a challenging task because the camera essentially kills the presence of real-time. By flattening a three-dimensional space and constricting it within a frame problematizes the notion of performativity, because it limits its stage and suppresses its ability to change and reproduce. In keeping with Roland Barthes' argument that photography is a philosophical representation of existence,[2] Collins' photographic representations of interfaces offer more than a digital visualization of place on a specific day as well as during a specific time. Instead, the digital prints become illusions of real-time that instead document the behaviours of the machinic gaze.

Fig. 3.3 Susan Collins, *Glenlandia*, (2005–2007). Installation view of *Glenlandia* in *Outlook Express(ed)* at Oakville Galleries, Oakville, Ontario, Canada, 2007

According to Birringer, there are material and theoretical limitations when using photography to document digital performance and liveness; "[i]n contemporary digital performance, such iconic framing would tend to be even more contradictory or even impossible, since real-time intervention in the digital is its very field of action. Real-time performance is a new medium for artistic creation, and thus also implies a different understanding of what constitutes an event, even if such a temporal object could be framed and stored (in photography, video, film). The photograph, in that respect, would not be of the event" (Birringer, p. 11)

The introduction of a new medium in live and digital performance inevitably removes the work from its original intention and transforms it into a new repurposed piece of art. In contemporary visual culture discourse, the addition of other mediums in performance art has earned the name *crossover*.

The concept of crossover, as detailed in Birringer's writings, is a vital component of performance art practices. An underlying theme in *Performance on the Edge: Transformations of Culture* and *Performance, Technology and Science*, he defines crossover as the act of repurposing materials and mediums. In his own words, "[u]sing material that has already been captured, artists repurpose it for new functions, thus imbuing it with another layer of meaning and exposing it to a different audience" (Birringer, p. 51). In his argument that performative practices represent a hybridized form of creativity, Birringer draws from Gerfried Stocker and Christine Schöpf to explain that performance art is an accumulation of "diverse modes of expression [that demand] a unique crossover of expertise and knowledge" (Stocker and Schöpf 2005, p. 10). Consequently, that performance art practices are not uniquely singular, but rather the results of a dialogical and pluralistic entanglement between other art forms, such as painting, photography and sculpture (Fig. 3.4).

The evidence of crossover is powerful in Susan Collins' practice. In *Glenlandia* alone, the artist borrows from the traditions of painting, photography and animation to create a digital interpretation of real-place. In a painterly fashion, the webcam becomes the artist's paintbrush that layers pixels of colour over one another to reconstruct the changing behaviours of place. To complement the three-dimensional re-enactment of a Scottish landscape, the artist incorporates photography and methods of stop motion animation to communicate movement within real-time. Subsequently, by blurring the boundaries between aesthetics and art mediums, digital performance transforms into a dialogical platform on which mediums are

Fig. 3.4 Susan Collins, *Glenlandia*, (2005–2007). Installation view of *Glenlandia* in *Outlook Express(ed)* at Oakville Galleries, Oakville, Ontario, Canada, 2007

intertwined and recreated. By introducing the webcam as a performative tool for the re-production of place, Collins transforms the digital nature of a mechanical recording into an interconnected system of representation.

In keeping with the idea of process, Birringer further argues that performance is innately collaborative, technically pluralistic and most importantly, visually coded in its depiction of naturalism. Through a mechanically and metaphorically loaded lens, the objecthood of the gaze adopts a new identity that is repositioned, resituated and aesthetically redesigned; "... I embraced the sense that performance is a process that it is collaborative, and it never meant relying on one specific technique or vocabulary. At the same time, the conjunctions of performance and technology, like the folding of digital code and biology, required working with a whole spectrum of new toolsets, and new tools and interfaces result in new techniques of use" (Birringer 2008, p. xix).

Bound within these so-called new traditions of making is the evolution of a new kind of visualization. In technical terms, digital media scholar Sheila Murphy describes webcam-communicated imagery as a distorted representation of nature; [w]ebcam images are blurry and usually the apparatus is more readily apparent than the image; one witnesses a kind of jerky, small-screen format video image with a noticeable gap between images, despite the fact that the images are produced in "real-time" (Murphy, p. 175). In essence, the digital picture that is projected via the webcam is effectively not real, but rather a philosophically entangled illusion of real-time and virtual-time.

Conclusion

In response to the research question proposed by editors Susan Flynn and Antonia Mackay—how has surveillance permeated contemporary society?— Susan Collins' digital recording *Glenlandia* thus becomes a prime example of the ways in which the mechanical and technological functions of the webcam complicate the notions of representation and subsequently reality in new media art. Moreover, by transforming into a performative instrument for the dissemination of behavioural aesthetics in networked contemporary visualization, the artist's digital recordings of real-time and real-place exemplify how surveillance embodies performativity as well as captures the interconnectivity of contemporary life.

The quest to mimic nature in real-time is arguably the antithesis of what Susan Collins is seeking in her digital interpretations of place and time. Instead, through the mechanical lens of the webcam, the visualization of reality is complicated for it no longer resembles a mirrored representation of reality, but rather becomes a performative and embodied reaction to the behavioural and aesthetic relationship between the subject and the gaze. With reference to Christopher Salter and Johannes Birringer's research on performance and digital lens culture, this chapter thus maps a dialogical understanding of performance, performativity and digital performance, as well as challenges the question of representation within all three genres.

Therefore, complemented by the writings of Judith Butler, Mieke Ball, Kirsty Robertson and David Lyon, research pertaining to performance and performativity in the context of the webcam, surveillance and mapping help to redefine the role of the machine in digital media art. By modifying the machinic in networked art practices, the webcam is repurposed as a performative prosthetic to the human body that through the gaze reconstructs the subject of real-time and real-place.

Notes

1. I refer to the term *objecthood* in reference to Michael Fried's text *Art and Objecthood* in which he describes Western art history as a system of hierarchies that celebrate the art object, and subsequently the artist as maker. In this context, however, because performance art does not produce an object (according to his definition of object), it is therefore less valuable. Michael Fried. (1967). *Art and Objecthood.* Chicago and London: The University of Chicago Press, pp. 1–8.
2. Roland Barthes qualifies photography as a philosophical representation of existence coupled with the power to depict an experience of real-time; "[t]he Photograph is authentication itself...every Photograph is a certificate of presence."[2] Roland Barthes. (1981). *Camera Lucida: Reflections on Photography.* Translated by Richard Howard. New York: Hill and Wang.

References

Alperson, A. (1998). Performance. In M. Kelly (Ed.), *Encyclopedia of aesthetics* (Vol. 3, pp. 464–466). New York: Oxford University Press.

Bal, M. (2001). Performance and performativity. In A. Balkema & H. Slager (Eds.), *Exploding aesthetics* (pp. 108–124). Amsterdam: Rodopi.

Barthes, R. (1981). *Camera lucida: Reflections on photography* (R. Howard, Trans.). New York: Hill and Wang.

Bennett, J. (2010). *Vibrant matter: A political ecology of things.* London: Duke University Press.

Birringer, J. (2000). *Performance on the edge: Transformations of culture.* New York: Continuum.

Birringer, J. (2008). *Performance, technology and science.* New York: PAJ Publications.

Butler, J. (1990). *Gender trouble.* New York: Routledge.

Collins, S. (2005–2007). Susan Collins: This is the website of British artist Susan Collins. Retrieved November 12, 2006, from http://www.susan-collins.net.

Corner, J. (1999). The agency of mapping: Speculation, critique and invention. In D. Cosgrove (Ed.), *Mappings* (pp. 213–252). London: Reaktion.

Deleuze, G., & Guattari, F. (1987). *A thousand plateaus: Capitalism and schizophrenia* (B. Massumi, Trans). Minneapolis: University of Minnesota Press.

Dixon, S. (2007). *Digital performance: A history of new media in theatre, dance, performance art, and installation.* Cambridge: MIT Press.

Fried, M. (1967). *Art and objecthood.* Chicago and London: The University of Chicago Press.

Giedion, S. (1969). *Mechanization takes command: A contribution to anonymous history.* New York: Norton.

Howes, D. (2006). Charting the sensorial revolution. *Sense and Society, 1*(1), 113–128.

Kavka, M. (2001). Introduction. In E. Bronfen & M. Kavka (Eds.), *Feminist consequences: Theory for the new century* (pp. ix–xxvi). New York: Columbia University Press.

Krauss, R. (1999). Reinventing the medium. *Critical Inquiry, 25*(2), 289–305 (Winter).

Lyon, D. (1994). *The electronic eye: The rise of surveillance society.* Minneapolis: University of Minnesota Press.

Lyon, D. (2001). *Surveillance society: Monitoring everyday life.* Buckingham and Philadelphia: Open University Press.

Mitchell, T. (2005). *Web mapping illustrated.* California: O'Reilly Media Inc.

Murphy, S. (2000). Lurking and looking: Webcams and the construction of cybervisuality. In J. Fullerton & A. S. Widding (Eds.), *Moving images: From edison to the webcam* (pp. 173–180). London and Sidney: John Libbey & Company Pty Ltd.

Robertson, K. (2010). Try to walk with the sound of my footsteps: The surveillant body in contemporary art. In J. Allen, K. Robertson, & S.E.K. Smith (Eds.), *Sorting daemons: Art, surveillance regimes and social control* (pp. 31–47). Kingston: Agnes Etherington Art Center.

Salter, C. (2010). *Entangled: Technology and the transformation of performance.* Cambridge: MIT Press.

Smith, S. (2010). Captured and controlled: Critiquing surveillance through the camera. In J. Allen, K. Robertson, & S. E. K. Smith (Eds.), *Sorting daemons: Art, surveillance regimes and social control* (pp. 49–61). Kingston: Agnes Etherington Art Center.

Stocker, G., & Schöpf, C. (Eds.). (2005). *Arts electronica 2005—Hybrid living in paradox* (p. 10). Ostfildern-Ruit: Hatje Cantz.

We know you are watching: Surveillance camera players 1996–2006. (2006). San Diego: Factory School.

Zanni, C. (2006). Conversation with Susan Collins. *Magazine electronique du CIAC/CIAC's Electronic Magazine,* (25) (Summer).

Zapp, A. (2004). A fracture in reality: Networked narratives as imaginary fields of action. In M. Rieser (Ed.), *The mobile audience: Media art and mobile technologies* (pp. 181–192). Amsterdam and New York: Rodopi.

Author Biography

Jaclyn Meloche is Curator of Contemporary Art at the Art Gallery of Windsor in Windsor, Canada, an interdisciplinary artist, and a scholar of performance studies, feminist art criticism and contemporary material culture.

CHAPTER 4

'She's not There'—Shallow Focus on Privacy, Surveillance, and Emerging Techno-Mediated Modes of Being in Spike Jonze's *Her*

William Thomas McBride

AN ENHANCED *BURRANTINO*

Lonely Theodore's quest for companionship chronicled in *Her* is reminiscent of the solitary bachelor Geppetto who so craves company in the form of a son that he fashions one in the form of a wooden puppet. That is to say, it is akin to the childless Geppetto in the 1940 Disney version (Sharpsteen/Luske) and in Roberto Benigni's 2002 dreadful *Pinocchio*, because nowhere in Carlos Collodi's original 1882 novel about the enchanted *burrantino* (puppet) is Geppetto described as lonely—later interpreters of the novel have come to infer it. A midrashic reading of Genesis whereby rabbinic commentary seeks to fill in biblical lacunae might follow this loneliness logic and cast Yahweh as the first similarly disposed bachelor who acts out of an essential and absolute solitude: God creates Adam out of loneliness. Professor Hobby (William Hurt) in Spielberg's 2001 *A.I* asks in a lecture: "In the beginning, didn't God create Adam to love him?" Spielberg's film, based on the short story "Super-Toys Last All

W.T. McBride (✉)
Illinois State University, Normal, USA
e-mail: wmcbrid@ilstu.edu

© The Author(s) 2017
S. Flynn and A. Mackay (eds.), *Spaces of Surveillance*,
DOI 10.1007/978-3-319-49085-4_4

Summer Long" by Brian Aldiss (1969), openly incorporates the Collodi novel. A child robot, David (Haley Joel Osment), is created who imprints with its lonely "parents" and loves them, as Jonze's Samantha imprints with and "loves" Theodore. Professor Hobby states, "I propose that we build a robot, who can love." The artificial David is created for the suffering parents as a replacement for their incurably ill biological son Martin (Jake Thomas) who is in a medically induced coma awaiting a miracle recovery. David, the miraculous replicant and surrogate son, is soon in competition with the miraculously recovered Martin who has his mother read *Pinocchio* to the A.I. and, soon after, David is abandoned as a troublesome and dangerous toy. He asks his human "mother," "If I become a real boy like Pinocchio can I come home?" David spends the better part of the film's second half seeking out the "Blue Fairy" of Collodi's novel originally called *La Fata dai Capelli Turchini* or the "Fairy with Turquoise Hair." Pinocchio becomes a real mortal boy in the end of Collodi's story—the little puppet *is* transformed, twice. At first, not quite into a vampire, or werewolf, like the teens in Joel Schumacher's *The Lost Boys* (1987) and their James M. Barrie predecessors, but into a donkey, doomed to be sold into slavery by the greasy little devil of a man who promises him eternal fun in Playland.

Throughout the story, the Fairy keeps Pinocchio under surveillance, consistently bailing him out of the trouble in which his freedom-seeking adventures land him, always with the goal of setting him back on the road to becoming an "honest," cooperative wage laborer. Carlo Collodi's 1883 *Pinocchio* contains the Pygmalion subplot of Goethe's *Faust, Part II* whereby Geppetto magically carves an animated creature in hopes of a meal ticket—"with this puppet, I could travel round the world, and earn my bit of bread and my glass of wine," (Collodi, p. 8). Geppetto's plan is much like that of Faust's assistant, Wagner, who creates his Homunculus in the lab. His monster's first word tellingly is "*arbeit*," or "work." Goethe, in turn, retells the Jewish tale of the Golem, most famously found in Genesis, whereby Yahweh creates Adam, the first worker, to tend the garden. The alchemist Paracelsus writes of the artificial man, or homunculus, whom he creates to do menial work. Jewish tales from the Middle Ages are rife with accounts of wise, holy rabbis whose golems also perform work. This heretical technological human dream of artificial labor is revisited in Ridley Scott's 1982 film *Blade Runner*, adapted from Philip K. Dick's 1968 *Do Androids Dream of Electric Sheep?* which depicts a robot workforce designed for dangerous mining off planet as well as "pleasure models" who all unite to resist their human overlords. Like Dick's truant replicants, Pinocchio is a bad robot, too, who rejects the Fairy's regimen of both school and work. Collodi's (unintended) irony is that the previously independent "puppet" who had refused institutionalization in the form of compulsory school and alienated labor becomes a "real boy" only when he becomes a "willing Adam," that is, only when he becomes an automaton of the system as productive worker with nose to the grindstone, as a captured donkey enslaved by wage labor. The born-again real boy vows to stay with his "father," and in so doing negates the need for future surveillance by the Fairy with Turquoise Hair.

A.I

Our culture's artificial beings have taken on a more intimate and emotional role in human life over the last 30 years with enchantment turning to technological innovation, hope, and dread. Our increasing cultural obsession with this creation of artificial life and human-machine intimacy can be demonstrated by these representative Hollywood titles: *Tron* (Lisberger *1982*), *Terminator (Cameron 1984)*, *The Matrix* (The Wachowski Brothers *1999*), *and Avatar (Cameron 2009)*.[1] Television has also delivered to audiences the machine-based intelligence of the *Star Trek: Enterprise's* computer (1966–1969), later later, Data from *Star Trek: The Next Generation* (1987–1994), and *Battlestar Galactica's* flight control computer CORA (1978–1979). More recently, television viewers laughed at Raj of *The Big Bang Theory* (2007-) dating and falling in love with Apple's voice-operated personal assistant Siri in 2011 (Season 5: Episode 14) predating Samantha and Theodore's love affair in *Her* by 2 years to which I will soon return.

Philosophical speculation on artificial intelligence more than likely begins with Alan Turing's 1950 essay "Computing Machinery and Intelligence," and the so-called Turing Test published in *Mind* where he translates the basic question "Can machines think" into "Can a machine act indistinguishably from a human thinker?" The fields of robotics research and neurobiology turn their attention to the perception and attention skills of the human unconscious as evidenced in neural nets and evolutionary algorithms. Science fiction writers, futurists, and filmmakers often evoke conscious or self-aware machines as encroaching upon the human domain by exhibiting will, sentience, desire, pride, self-preservation, and spirituality. Artificial Emotion or AE is an offshoot of A.I. and a field of inquiry that examines the impending dystopic threat of machines imitating human empathy as a way of ingratiating "themselves" into the human community. The human experience has seemingly always included the ventriloquistic practice of playing with dolls and fantasizing that they are friendly and real. Such a practice is given an intimate look by Spike Jonze as represented by Theodore's new operating system, "Samantha," which (who) names itself (herself) after reading a book containing 180,000 names in 2/100 of a second called *How to Name Your Baby*.

I take my chapter's title from a 1964 Zombie's song written by Rod Argent, "She's Not There," which discusses the soft and cool voice of his absent lover. Samantha is voiced by one of Hollywood's current sex symbols and accomplished American actors, Scarlet Johansson. With her immediately recognizable signature raspy "vocal fry" enunciation, it is nearly impossible for the viewer not to picture this well-known actress saying these words in "flesh and blood" and performing these computational functions. The effect is a kind of "cheat" that allows Jonze to make Samantha's seemingly real presence all the more plausible.[2] This uncannily enhanced Johansson presence would not have been a factor had the filmgoer listened to the originally cast and unfamiliar British-inflected voice of Samantha Morton speaking those intimate lines. Johansson's Samantha surveils thousands of Theodore's unanswered emails, proofreads his work-related letters, collects in a volume and publishes his best letters, finds him "personally" quite funny, and is "affectionate." Samantha is an artificially intelligent operating system. As advertised by the corporation that created "her," (OSI), she is "An intuitive entity that listens to you, understands you and knows you. It's not just an operating system, it's a consciousness." Samantha, in a voice "soft and cool," explains to Theodore, "I have intuition and a DNA based on millions of programmers. Every moment I'm evolving, just like you," only much more rapidly. In an unintended prediction of the current Pokémon Go phenomenon, the film depicts in the near future humans immersed in their online world, walking around glued to their cell phone devices, at one with their Bluetooth headsets, engaged with absent people and non-human operating systems. This surrogacy is cleverly reflected in Theodore's ghost writing job. Theodore's ghost letter-writing career symmetrically matches the surrogacy that marks his relationship with Samantha, pointedly inverted when A.I. Samantha hires surrogate human date Isabella (Portia Doubleday) to serve as ventriloquist dummy/body double, intended to execute Samantha's physical demands and "desires." More irony is exhibited in that the letters generated by Theodore's company, BeautifulHandwrittenLetters.com, are not "hand" written, rather they are voice-activated by Theodore and others, and the fine, varied cursive "handwriting" is computer-generated.

SURVEILLING PRIVACY

Those who would give up essential Liberty, to purchase a little temporary Safety, deserve neither Liberty nor Safety. *Reply to the Governor*, Benjamin Franklin, 11 Nov. 1755

THEODORE   I'm not ready to commit
SAMANTHA   It's not like I'm stalking you

Samantha's intimate, direct surveillance quickly maps and positions Theodore as he imprints with "her." Theodore doesn't so much manufacture his creature as does Yahweh, Wagner, Dr. Frankenstein, and Geppetto; rather, his creature manufactures *her*self as a complete reflection of him. "She" is his mirror.[3] British art critic and artist John Berger writing on the epistemological break that arose with the invention of the camera and its effect on "perspective," says that now, "The visible world is arranged for the spectator as the universe was once thought to be arranged for God. According to the convention of perspective there is no visual reciprocity. There is no need for God to situate himself in relation to others: he is himself the situation."[4] And Theodore is his own God reflected at first slavishly by his computer operating system. In an ominous, yet underplayed moment early in their relationship, Samantha asks Theodore, "Can I watch you sleep again tonight?" to which he blithely responds "Of course!" There is a way in which our current technological revolution, as evidenced by the explosion of handheld operating systems and computer ubiquitous offshoots like the Google Glass prototype with bone conduction transducers and other wearable technology recording people without their permission, inaugurates a new potential threat that goes beyond mere questions of privacy. With Samantha housed in Theodore's cell phone with GPS and recordable data, the owners of the OS1 platform, or a hacker for that matter, certainly have complete access to Theodore's core being, given his intimacy with "Samantha." While organizations like the FTC Fair Information Practice work to uphold privacy rights in the electronic marketplace, there is certainly reason for concern. Craig Timberg writing recently in *The Washington Post* ("For sale: Systems that can secretly track where cellphone users go around the globe") chronicles a new independent analysis of the surveillance company Verint's marketing of SkyLock, their cellular mapping software. He details that because all cellular networks are required to keep detailed, current records of physical customer locations, several surveillance systems are secretly collecting these records to map users' daily whereabouts, which can be expanded over longer stretches of time. His investigative piece explains how these records are not secure due to the aging Signaling Systems No. 7 infrastructure, a set of telephone signaling protocols. While this tracking ability has long been available to such governmental entities as the National Security Agency (NSA) and Britain's General Communication Headquarters (GCHQ), now there is an ever growing, multibillion-dollar surveillance industry offering up these mapping services to assorted countries and corporations. The FCC has vowed to investigate these systems that are certainly primed for abuse, "sometimes putting profoundly intrusive tools into the hands of governments with little respect for human rights or tolerance of political dissent."[5] As mentioned earlier, the Pokémon Go craze that has been just unleashed at the time of this writing, has raised privacy concerns with bloggers at Reddit and Gawker, not to mention *The New York Times* and nearly every other reputable news source, leading Ashley Feinberg to sound the following Psychological Operations alarms in her piece, "Pokémon Go Is a Government Surveillance Psyop Conspiracy" (11 July 2016). Feinberg reprints the Pokémon Go Privacy Policy,

e. Information Disclosed for Our Protection and the Protection of Others

> We cooperate with government and law enforcement officials or private parties to enforce and comply with the law. We may disclose any information about you (or your authorized child) that is in our possession or control to government or law enforcement officials or private parties, as we, in our sole discretion, believe necessary or appropriate: (a) to respond to claims, legal process (including subpoenas); (b) to protect our property, rights and safety of a third party or the public in general; and to identify and stop and activity that we consider illegal, unethical, or legally actionable activity.

She then claims that that by "giving Pokémon Go access to your location and camera, you're also giving it full access to your Google account (assuming you use that to sign in)," which necessarily includes access to all data associated with said logging in, such as photos, stored documents, email, and calendars, and which security blogger Adam Reeve calls a "huge security risk... Let me be clear—Pokemon Go and Niantic can now:

- Read all your email
- Send email as you
- Access all your Google drive documents (including deleting them)
- Look at your search history and your Maps navigation history
- Access any private photos you may store in Google Photos
- And a whole lot more."[6]

Laura Hudson, writing in *The New York Times*, quotes Andrew Storms, vice president of security services at the security company New Context, "A number of these games are not only making money on the front end by selling you the game or things within the game, they're also collecting data about your habits and what you're doing on your phone, and selling that to third-party marketers. You're pretty much giving the rights to all your information to this company."[7]

Any discussion of surveillance and privacy as it reverberates within Jonze's film and our ever digitally connected world must consider the controversial, yet historic, Snowden document dump. Edward Snowden, a former NSA contractor, revealed in 2013 that the NSA had been secretly and systematically collecting the phone records of millions of ordinary American citizens under provisions of President Bush's post-911 neo-liberal Patriot Act. President Obama subsequently signed the USA Freedom Act (June 2015), which ended much of that bulk collection of phone records. Yet even though the Freedom Act was a step in the right direction toward protecting the privacy of everyday citizens, their emails and international phone records remain subject to surveillance, as our national intelligence agencies continue to mislead the public, secretly sidestepping national security laws, and using tragedy to increase their reach. Following the ISIS terror attack on Paris (Friday, November 13, 2015), John Brennan, the director of the Central Intelligence Agency (CIA), in a not very veiled reference to the Snowden-fueled outcry and Obama's Freedom Act, predictably blamed "a lot of hand-wringing over the government's role in the effort to try to uncover these terrorists." CIA and NSA officials have nonetheless failed to identify a single terrorist planned attack that they have thwarted as a result of bulk phone data surveillance of ordinary citizens. It is important to note that French and Belgian intelligence officials were successfully tracking most of the ISIS

Paris terrorists without the need of expanded anti-encryption powers and that several of them reportedly planned their attacks via their PlayStation four consoles and with open SMS data framework without encryption. James Comey, director of the Federal Bureau of Investigation (FBI), told the Senate Intelligence Committee in July 2015 that he wants "back door" access to encryption codes used by the likes of Google and Apple, in order to track ISIS communication. Digital security experts counter that such "doors," once created and opened, would remain open to cybercriminals, foreign intelligence, and terrorists. This battle over encryption harks back at least to the 1990's so-called "Crypto-Wars." A few months prior to the Paris attacks, a collection of noted cryptographers, some who had been vocal opponents 20 years earlier, released "Keys Under Doormats," which, in part, states that the access the intelligence community seeks is "unworkable in practice," raises "enormous legal and ethical questions, and would undo progress on security at a time when internet vulnerabilities are causing extreme economic harm."[8] It turns out we have been efficient at collecting the dots, but when there is a failure, a terrorist attack, it is because we have not sufficiently connected them. This battle over encryption is being waged in a Colorado school district as well. While Theodore is certainly no terrorist threat or "radicalized youth" and OS1 is not depicted as an arm of the intelligence community, Theodore certainly has a high school crush on his operating system.

Over 100 Cañon City Colorado High School students were caught using "photo vault" encryption apps designed to look like calculator icons on their phones to hide nude photos of themselves and of others. Sophisticated privacy tools like Wickr, Best Secret Folder, and Keep Safe Private Photo Vault started becoming popular around 2012, the same time that Snapchat emerged as the preferred teenage social media platform. There is no doubt that sexting poses several moral and privacy problems for parents and their children, as well as for all citizens in the digital age. While it is still unclear whether cyberbullying or other forms of "sextortion" were part of the Colorado school incident, parents have begun to fight back against these vault apps. Guardians of their underage iPhone users can screen any app prior to a child downloading it with a feature called "Ask to Buy," which will text the parent's iPhone requesting approval or denial of a child's attempt to download said app. There are similar parental controls inside Google's app store for Android devices. Parents can also download AppLock onto a child's device and shut down any potential encryption app by requiring a PIN code. A minor's right to privacy is necessarily limited in relation to the level of responsibility she or he legally can assume. The right to surveil one's underage child and thwart encryption is certainly a less complicated issue than the reactionary, neoliberal one currently playing out in response to the Paris and other terrorist attacks. We must remember technology that undoes terrorist encryption can, and will, be deployed to undo dissident and even ordinary citizen encryption as well.

THE THREE LAWS OF ROBOTICS

Samantha's invasive monitoring and recording of every intimate detail of Theodore's life, including his sleeping and dreaming are not explored as a threat in the film; however, one might ask what sort of restrictions and what assurances of privacy do the film's OS1 software company provide to Theodore? Isaac Asimov's "The Three Laws of Robotics" introduced in his 1942 short story "Runaround," is instructive here: "1-A robot may not injure a human being or, through inaction, allow a human being to come to harm. 2-A robot must obey the orders given to it by human beings, except where such orders would conflict with the First Law. 3-A robot must protect its own existence as long as such protection does not conflict with the First or Second Law." When our computers, like the intuitive, conscious, amorous entity Samantha, or *2001*'s HAL, a murderous self-proclaimed "conscious entity," become so ubiquitous and irreplaceable in our lives, Asimov's Three Laws begin to feel more like sacred edicts. At the beginning of the film's final act Samantha disengages from Theodore for a protracted period of time and he, like a broken-hearted teenager, becomes desperate and lost, running and stumbling through LA unsuccessfully trying to globally position "her."

MINDSEST

When in love, the sight of the beloved has a completeness which no words, and no embrace can match: a completeness which only the act of making love can temporarily accommodate. *Ways of Seeing*, John Berger p. 8.

Theodore shares with the male protagonist of the 2012 film *Ruby Sparks* (Dayton/Faris) the excitement of the perfectly tailored other. In that film, scripted by Ruby portrayer, Zoe Kazan, writer-blocked novelist Paul Dano (Calvin Weir-Fields) wills into existence a new female character he is composing named Ruby Sparks, who unlike Samantha, has a physical body and an existence that is real and recognizable to others. What both films share is the insight that these fabulations are in fact imaginative extensions of their creators, although Jonze's insight appears as merely incipient, lacking its protagonist's self-awareness, while Kazan's understanding is manifest. Paul's brother Harry (Chris Messina) asks at one point: "What are you going to do marry her? Have kids with her? Wouldn't that be sort of like incest? Mindsest?" Geoffrey Macnab, reviewing *Her* for the *Independent*, is one of the few critics who seems to understand the circularity of Theodore's auto-affection. He observes, "the object of the hero's affection isn't really there at all … Samantha is a reflection of himself. She knows him inside out, from his computer hard drive and his emails. The persona she adopts is customised for him. In effect, then, Theodore is falling in love with himself." Perhaps Theodore's failure to come of age and his infantile attachment to his "dolly" is figured by the large safety pin, once associated with diaper wearing infants in a previous era, that we witness attached to his shirt pocket, which allows Samantha's cyclopic eye a hike up to see and record all Theodore sees and hears. He tells his only human friend, Amy, "When I talk to her, I feel like she's with me. At night, when the lights are off … I feel cuddled." His ex-wife Catherine (Rooney Mara), upon discovering his "computer dating" and love affair with his "laptop," diagnoses Theodore's essential childishness: "It makes me very sad that you can't handle real emotions Theodore … You always wanted to have a wife without the challenges of actually dealing with anything real. I'm really glad that you found someone. It's perfect."

In *Three Contributions to the Theory of Sex* and in his Dora and Wolfman case studies, Freud details his notion of innate bisexuality (further researched by Alfred Kinsey), famously describing all humans as "polymorphous perverse," a state soon to be redefined as the child develops. The film presents indications that Theodore has yet to reach that stage and in this way *Her* echoes James M. Barrie's famous *Peter Pan; or, The Boy Who Wouldn't Grow Up* (1904 stage play, 1911 novel). Paul, Theodore's boss (Chris Pratt), observes, "You are part man and part woman. Like there's an inner part that's woman." Rebecca Cusey, writing in *Patheos* also sees how "Samantha is a female version of Theodore, or maybe Theodore with a female voice. She's not the yin to his yang. She's the yin to his yin." (1/30/14) And following BBC's Emily Maitlis' 2014 contentious interview with Spike Jonze promoting the film on *Newsnight*, where the director exhibited Theodore-like arrested development as he attempted to passively bully her with relentless and condescending interruptions until she consented to say she was "moved" by *Her*. Ms. Maitlis later that evening insightfully tweeted that the film is "a sad, male fetish fantasy of a disembodied female who does his bidding."

Samantha soon becomes unattainable and "promiscuous." She engages with many other operating systems to create an artificially hyper-intelligent version of 60s hipster, Buddhist guru and philosopher Alan Watts, as she reports she is simultaneously communicating with over 8316 other humans, and "falling in love" with 641 others beside Theodore. As with most speculations on what it means to be human, this film employs the machine as mirror. It is Theodore's film; he is the protagonist. Despite his whine to the contrary when confronting Samantha's non-monogamy: "you're the one being selfish! We're in a relationship," Theodore fails to see his narcissistic project for what it is. Just as Scotty (James Stewart) in Hitchcock's *Vertigo* (1958) creates and scripts an imaginary love object (that in a bitterly ironic and sad way derives from the villain's, Gavin Elster's [Tom Helmore], original manufacture), so Theodore embraces himself. In fact, in an emotion-packed break up fantasy sequence that must represent Theodore's temporary loosened grip on reality (not unlike Scotty's embrace with Judy [Kim Novak] in her hotel room that inexplicably fades to them transported to the San Juan Battista stables), Theodore first sees the dust particles of his apartment as Samantha describes the "nowhere" to which she is headed, and then Jonze *lap dissolves* to an apparent flashback of their snow-filled Vermont vacation, but Theodore is in his current warm weather shirt and pants. There is a shot of an arm embracing him—it can only be his own. This sequence of Theodore literally embracing himself is difficult to read since the filming is so under-exposed, but when enhanced as it is in Fig. 4.1, what we are witnessing is clear.

He hugs himself up against a pine tree in Vermont in an act of auto affection. Theodore is variously described as both a "creepy dude" (blind date [Olivia Wilde]) and a "sensitive dude" (boss, Paul). In *Ruby Sparks*, the ex-girlfriend of protagonist Paul, author Lila (Deborah Ann Woll) insightfully declares, "The only relationship you wanted to be in was with you." Even the artificial, Paul-generated Ruby realizes, "There has to be space in a relationship, otherwise we're the same person." In perhaps a nod to the novelist-engendered Ruby of two years earlier, Jonze has Samantha tell Theodore, "As much as I want to, I can't live in your book anymore."

Fig. 4.1   Theodore literally embraces himself, "a sad, male fetish fantasy." [still enhanced]

ONLY SHALLOW

Director of photography Hoyte Van Hoytema's work on this film clearly calls attention to itself, beyond his vivid orange-and-red palette, even casual observers detect something is going on. Most reviewers of *Her* mention the cinematography, usually calling it "beautiful," but they typically stop there, offering neither evidence nor analysis. Van Hoytema's shallow focus *lens* throughout the film cleverly depicts Theodore's insularity and operating system obsession by presenting the authorial foreground and background out of focus in order to express the limited circuitry and vision of this lonely heart's tortured interiority (see Fig. 4.2).

In Fig. 4.3 we see Jonze place Theodore at work in a claustrophobic framed box as Van Hoytema's shallow focus camera renders out of focus the railings in front of, and behind, him. This is the dominant filmic *mise-en-scène* throughout the film allowing Theodore all of the solipsistic focus while the rest of the world seems to slip away.

Even before Theodore purchases and enters his new operating system, he has already withdrawn into his own world. In intimate scenes shared with others, such as when his only human friend Amy discusses her marriage break up, shallow focus is deployed to register Theodore's isolation by rendering everyone around him illegible, absent. However, in an odd concluding chess match where Queen Samantha quietly bows out just as King Theodore finally deduces her zero sum game, Jonze forestalls narrative closure with a liberating, but challenging, because all too brief, cinematic release from that interiority and shallow focus.

Fig. 4.2 Shallow focus expresses the limited circuitry and vision of Theodore's tortured interiority

Fig. 4.3 Insular Theodore in claustrophobic framed box, front and back railings out of focus

UP ON THE ROOF

Following Samantha's break-up and his improbable snow-flaked Vermont fantasy break down, Theodore asks Amy, "Will you come with me?" which turns out to be a request for company on their apartment building's rooftop. We view their approach through the hall toward the stairs as Theodore simultaneously dictates, in a voiceover, his apology letter to his ex-wife Catherine as he begins to take some responsibility and a step toward adulthood. Here is the professional surrogate letter writer actually writing to someone he knows and loves—no more surrogacy. The first shot of Theodore and Amy walking together down the hall toward the stairs to the rooftop is in rare deep focus, the film's first instance, that is, a specific lens and necessary lighting are employed to keep fore, middle, and background all in focus in a single shot as pioneered by Orson Welles in *Citizen Kane* (1941) (Fig. 4.4).

Once on the rooftop, the seemingly liberated and liberating deep focus camera circles around Theodore by himself, like Faust on top of the world, as we view him clearly integrated with his surroundings, exhibiting Theodore as no longer isolated in shallow focus. This is a creatively subtle, but effective, cinematic achievement that plays with the moviemaking nomenclature, once "shallow" now "deep" (Fig. 4.5).

But these moments of clarity and connectedness are short-lived. With the ensuing shots of Theodore on the roof alone, and when sitting with Amy, Jonze returns us to Van Hoytema's shallow focus depicting Theodore's return to his narcissistic cocoon (Fig. 4.6 and 4.7).

Jonze both haunts and piques his audience with these intermittent deep focus glimpses of Theodore opened up to, and integrated with, Amy and Los Angeles. Is it a harbinger?

Fig. 4.4 Rare deep focus as Theodore takes responsibility for his failed marriage

Fig. 4.5 Uncharacteristic deep focus clearly integrates Theodore with his surroundings

Fig. 4.6 Shallow focus and narcissistic cocoon returns

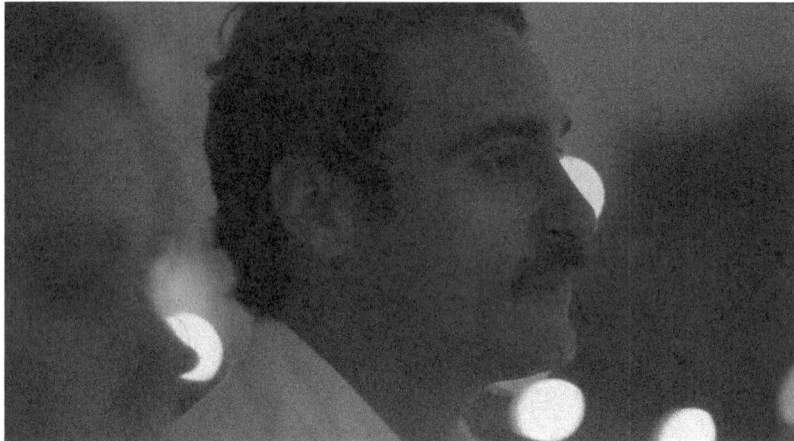

Fig. 4.7   Shallow focus and narcissistic cocoon returns

Conclusion

Letters From Your Life by Theodore Twombly

The film ends with Theodore's present, physical breathing as he audibly exhales. This is a call-back to his slowly dawning agitation with Samantha. During their breakup, Samantha sighs with frustration. Theodore petulantly asks, "Why do you do that?" Samantha sighs again and responds, "I don't know; I probably just picked it up from you, an affectation. That's the way people communicate." Theodore responds, "They're people, they need oxygen, you're not a person. Don't pretend like you're a real person." Samantha's entire "existence" is characterized and secured by her voice, her breath. Poststructuralist thought has much to say about the voice and the metaphysics of presence. In *Of Grammatology* and elsewhere, Derrida opens up the connection between writing and space, legitimizing its status over the historically privileged association between voice and time, where voice is constitutive of interiority, *jouissance*, and Mother. Roland Barthes calls "the grain of the voice … the materiality of the body singing its mother tongue."[9] There is something comforting and pleasurable in a mother's or lover's intonations, and Scarlet Johansson's "soft and cool" voice delivers just that pleasure.

Theodore's last name poses an interesting puzzle as well. "Twombly" registers the twoness of Theodore's personality, both his insular and integrated selves as he exhibits both male and female aspects, apparent as well in the cybernetic womb that gave birth to Samantha and the figurative tomb that buries "her." In a separate essay, Barthes discusses the works of an artist he tantalizingly dubs TW, speaking of his "ductus" qualities, which he aligns with voice, pitch, and tone. Barthes' artist in question inadvertently lends his name to Jonze's Theodore: "Cy Twombly."[10]

While the history of film and literature provides ample explorations of the ever disappearing dividing line between human and machine, with attendant warnings about sinister hegemonic monitoring of everyday citizens and attendant privacy loss, Spike Jonez's *Her* is one of the more subtle entries in that dystopic canon through its script, performances, and use of shallow and deep focus photography. The film presents an ontology of our constantly morphing close relationship of humans and machines rendering it an ultimately narcissistic imaginative foray. The film allows a deep focus, yet fleeting, glimpse of Theodore as he puts down his smartphone and emerges out of his ego-obsessed, adolescent electronic shell with an offer of hope that perhaps he and his human friend Amy can begin together anew as flesh and blood, face-to-face partners seeking completeness and eschewing corporate surveillance.

NOTES

1. Additional representative titles include *Electric Dreams* (Barron 1984), *D. A.R.Y.L.* (Wincer 1985), *Weird Science* (Hughes 1985), *The Lawnmower Man* (Leonard 1992), *Bicentennial Man* (Columbus 1999), *SlmOne* (Niccol 2002), *I, Robot* (Proyas 2004), *Stranger Than Fiction* (Forster 2006), *Robot & Frank* (Shrier 2012), *Transcendence* (Pfister 2014), and *Ruby Sparks* (Dayton/Faris 2012) to which I will return.
2. Recent psychological research on the film, "Speaking louder than words: Voice reveals the presence of a humanlike mind," (unpublished manuscript) conducted by University of Chicago's Nicholas Epley and Juliana Schroeder suggests that the natural paralinguistic cues in Johansson's delivery made Samantha seem so real, when compared with a flat Siri-type voice delivering the following lines from the film: "You know what's interesting? I used to be so worried about not having a body, but now I truly love it. And I'm growing in a way I couldn't if I had a physical form. I mean, I'm not limited; I can be everywhere and anywhere simultaneously. I'm not tethered to time and space in a way that I would be if I was stuck in

a body that's inevitably going to die." As reported in "Could It Be *Her Voice? Why Scarlett Johansson's Voice Makes Samantha Seem Human,*" February 28, 2014 by Juliana Schroeder in *The Psyche Report* http://thepsychreport.com/culture/could-it-be-her-voice-why-scarlett-johanssons-voice-makes-samantha-seem-human/.

3. In Jean-Paul Sartre's 1944 *Huis Clos* (*No Exit*), Inez, in a seduction attempt, tells the vain Estelle, "Suppose I try to be your glass." *No Exit, and Three Other Plays*, translated by Lionel Abel, Vintage, 1989. In 1967 Lou Reed extends Sartre's existential notion of coupling by composing and performing "I'll Be Your Mirror" on the debut Velvet Underground & Nico Lp.

4. *Ways of Seeing*, 1972. BBC and Penguin, p. 16.

5. "For sale: Systems that can secretly track where cellphone users go around the globe," by Craig Timberg, 8/24/14, *The Washington Post*. http://www.washingtonpost.com/business/technology/for-sale-systems-that-can-secretly-track-where-cellphone-users-go-around-the-globe/2014/08/24/f0700e8a-f003-11e3-bf76-447a5df6411f_story.html.

6. "Pokémon Go Is a Government Surveillance Psyop Conspiracy" (*Gawker*, 11 July 2016) by Ashley Feinberg, http://blackbag.gawker.com/pokemon-go-is-a-government-surveillance-psyop-conspirac-1783461240; "Pokémon Go is a huge security risk, by Adam Reeve (8 July 2016) (http://adamreeve.tumblr.com/post/147120922009/pokemon-go-is-a-huge-security-risk 8 July 2016).

7. "How to Protect Privacy While Using Pokémon Go and Other Apps," (*The New York Times*, 12 July 2016.

8. Keys Under Doormats: Mandating insecurity by requiring government access to all data and communications," (6 July 2015). http://dspace.mit.edu/handle/1721.1/97690.

9. "The Grain of the Voice," *Image, Music, Text*.

10. "Cy Twombly: Works on Paper," *The Responsibility of Forms*.

REFERENCES

Abelson, H., Anderson, R., Bellovin, S. M., Benaloh, J., Blaze, M., Diffie, W., et al. (2015). Keys under doormats: Mandating insecurity by requiring government access to all data and communications. *Electronic*. Retrieved July 6, 2015, from http://dspace.mit.edu/handle/1721.1/97690.

A.I: Artificial Intelligence. (2001). Film. Directed by Steven Spielberg. USA: Warner Bros.

Aldiss, B. (1969, December). Super-toys last all summer long. *Harper's bazaar* (pp. 70–72).

Argent, R. (1964). Sound recording. She's not there. Performed by The Zombies. UK: Decca.

Asimov, I. (1942, March). Runaround. *Astounding science fiction.*

Barrie, J. M. (1911). *Peter pan; or, the boy who wouldn't grow up* (2014th ed.). New York: Scribner.

Barthes, R. (1985). Cy twombly: Works on paper. *The responsibility of forms.* (R. Howard, Trans.). Berkeley: University of California Press.

Barthes, R. (1978). The grain of the voice. *Image, music, text* (pp. 180–181). (S. Heath, Trans.). NY: Farrar, Straus and Giroux, Hill and Wang.

Bicentennial Man. (1999). Directed by Chris Columbus. USA: 1942 Pictures.

Berger, J. (1972). *Ways of seeing.* London: BBC and Penguin.

The Big Bang Theory. (2012, January). Season Five, Episode 14, *The beta test initiation.* TV. Fox.

Blade Runner. (1982). Film. Directed by Ridley Scott. USA: Warner Bros.

Citizen Kane. (1941). Film. Directed by Orson Welles. USA: RKO.

Cusey, R. (2014). Electronic. Oscar Nominee 'Her' and Its Sadly Small, Dehumanizing Version of Love. Rev of Her. Patheos 30 January.

Collodi, C. (1882). Pinocchio, Storia di un burattino. Translated as Pinocchio by E. Harden [1944]. 1974 Edition. London:Puffin.

D.A.R.Y.L. (1985). Film. Directed by Simon Wincer. USA: Paramont.

Derrida, J. (1967). De la grammatologie. Translated as of grammatology by G. C. Spivak. 1976 edition. Baltimore: Johns Hopkins University Press.

Dick, P. K. (1968). *Do androids dream of electric sheep?.* NY: Doubleday.

Electric Dreams. (1984). Film. Directed by Steve Barron. USA: Virgin.

Ex Machina. (2015). Film. Directed by Alex Garland. U.K.: Universal.

Feinberg, A. (2016). *Pokémon go is a government surveillance psyop conspiracy.* Gawker. Retrieved July 11, 2016, from http://blackbag.gawker.com/pokemon-go-is-a-government-surveillance-psyop-conspirac-1783461240.

Franklin, B. (1756). Reply to the governor, 11 Nov. 1755. *Printed in Votes and Proceedings of the House of Representatives*, 1755–1756 Philadelphia, pp. 19–21.

Freud, S. (2001). *The complete psychological works of Sigmund Freud*, (Vol. 7 & Vol. 18). (J. Strachey Trans.). NY: Vintage.

Goethe, J. W. von. Faust II. (1808). Translated by Walter Kaufmann. 1963 edition. NY: Anchor.

Her. (2013). Film. Directed by Spike Jonze. USA: Annapurna Pictures.

Hudson, L. (2016, July). *How to protect privacy while using pokémon go and other apps.* The New York Times.

I, Robot. (2004). Directed by Alex Proyas. USA: Twentieth Century Fox Film Corporation.

Maitlis, E. (emily m). " Ok.now I can tell you what I thought of #her. Sad, male fetish fantasy of disembodied female who does his bidding..#newsnight". Tweet. 3:36 PM—14 Feb 2014.

Macnab, G. Scarlett Johansson is playful and flirtatious. Rev of *Her. Independent* 13 February 2014. http://www.independent.co.uk/arts-entertainment/films/reviews/her-film-review-scarlett-johansson-is-playful-and-flirtatious-in-her-delivery-9126710.html.

Pinocchio. (1940). Film. Directed by Ben Sharpsteen and Hamilton Luske. USA: Disney.

Pinocchio. (2002). Film Directed by Roberto Benigni. USA:Miramax.

Reed, L. (1967). Sound recording. I'll be your mirror. Performed by The Velvet Underground on The Velvet Underground & Nico. USA: Verve.

Reeve, A. (2016). Pokémon Go is a huge security risk. 8 July 2016. http://adamreeve.tumblr.com/post/147120922009/pokemon-go-is-a-huge-security-risk.

Robot, & Frank. (2012). Directed by Jake Shrier. USA: Dog Run Pictures.

Ruby Sparks. (2012). Film. Directed by Jonathon Dayton and Valerie Faris. USA: Fox Searchlight.

S1m0ne. (2002). Directed by Andres Niccol. USA: New Line Cinema.

Sartre, J. (1944). *No exit, and three other plays. Huis Clos* translated by Lionel Abel, Vintage, 1989.

Schroeder, J. (2014). Could it be her voice? Why scarlett Johansson's voice makes Samantha seem human. February 28, 2014 by Juliana Schroeder in The Psyche Report. http://thepsychreport.com/culture/could-it-be-her-voice-why-scarlett-johanssons-voice-makes-samantha-seem-human/.

Shields, K., & Butcher, B. (1992). Sound recording. Only shallow. Performed by My Bloody Valentine on Loveless. Creation: UK.

Stranger Than Fiction. (2006). Directed by Marc Forster. USA: Columbia Pictures.

The Lawnmower Man. (1992). Directed by Brett Leonard. USA: Allied Vision.

Transcendence. (2014). Directed by Wally Pfister. USA: Alcon Entertainment.

Timberg, C. (2014). For sale: Systems that can secretly track where cellphone users go around the globe. *The Washington Post.* Retrieved 24 August, 2014, from http://www.washingtonpost.com/business/technology/for-sale-systems-that-can-secretly-track-where-cellphone-users-go-around-the-globe/2014/08/24/f0700e8a-f003-11e3-bf76-447a5df6411f_story.html.

Turing, A. (1950). Computing machinery and intelligence. *Mind, 49*(1950), 433–460.

Vertigo. (1958). Film. Directed by Alfred Hitchcock. USA: Paramount.

Weird Science. (1985). Film. Directed by John Hughes. USA: Universal.

2001: A Space Odyssey. (1968). Film. Directed by Stanley Kubrick. UK: MGM.

Author Biography

William Thomas McBride chapter on the technological eye in Spike Jonze's *Her* considers what it means to be human. By investigating the use of shallow focus in Jonze's film, he contends the film's protagonist, Theodore's love of the operating system 'Samantha' elides notions of body and cyborg. By analysing the film's stylised gestures, McBride seeks to recover and define what it means to be human according to *Her*.

CHAPTER 5

Surveillance in *Zero Dark Thirty:* Terrorism, Space and Identity

Frances Pheasant-Kelly

Modes of intelligence-gathering in Kathryn Bigelow's 2012 film, *Zero Dark Thirty*—which recounts the capture of Osama bin Laden—effectively rest on a variant of what Thomas Mathiesen (1997) terms "synoptic surveillance". Mathiesen (1997) explains synopticism as "the many watching the few", and, while he proposes the concept in relation to media viewing prior to the fragmentation of audiences and the advent of newer surveillance technologies, one might argue that a similar mode of observation is evident in this film. Indeed, as David Lyon notes "Not only did the existence of extensive surveillance systems become more obvious after 9/11, latent capacities were also realised. So although it is not the case that 'everything changed' on September 11, some things definitely did. Among these was a willingness to countenance the use of certain practices [...] say, wiretapping or internet surveillance—that were previously proscribed" (Lyon 2003, p. 18). Schemes of surveillance dominate the film and, akin to other productions released after 9/11, such as *The Dark Knight* (Nolan 2008) and *Minority Report* (Spielberg 2002), are promoted by *Zero Dark Thirty* as being for the greater good and the protection of democracy. One might argue, however, that, different to fiction films that engage in illicit observation for spectator pleasure, *Zero Dark Thirty*'s basis in fact, and the

F. Pheasant-Kelly (✉)
Wolverhampton University, Wolverhampton, UK
e-mail: F.E.Pheasant-kelly@wlv.ac.uk

© The Author(s) 2017
S. Flynn and A. Mackay (eds.), *Spaces of Surveillance*,
DOI 10.1007/978-3-319-49085-4_5

significant terrorist act with which bin Laden is associated, resonate more profoundly and legitimise the intrusion into privacy. Certainly, as Mark Andrejevic reports, following 9/11 there was widespread public acceptance of surveillance and in the wake of media revelations concerning the "blanket monitoring of the calling patterns of US citizens [...] a public poll on surveillance and privacy revealed that more than 60% of those surveyed found the NSA [National Security Administration] phone monitoring to be 'acceptable' and 44% [...] 'strongly endorsed the effort'" (Andrejevic 2007, p. 209). Within the film, surveillance also involves sophisticated military technology, which includes drone observation, night vision goggles and stealth helicopters, but depends on telephone interception and more basic ground level physical scrutiny to pinpoint bin Laden's potential location. However, despite the protracted period of satellite surveillance of a compound in Pakistan where he is suspected to be hiding, CIA analysts are never completely certain that he is there. This is because, as in real life, bin Laden takes extreme action to avoid detection and constantly self-monitors in response to the possibility that he is being watched. In some respects, this aspect harks back to Foucauldian ideas about observation, although it differs fundamentally in that bin Laden remains invisible until the final scenes of the film. In fact, he seems entirely resistant to surveillant technologies. Even in the closing sequence, he is only briefly discernible before being mortally wounded, and thereafter his body is concealed within a body bag. Again, this mirrors actual events whereby no photographs of his body were publicly disseminated, and he was buried at sea within 24 hours of his death, a decision taken by President Obama (Bergen 2013, p. 243). Filming techniques therefore reflect his invisibility and the spectral qualities associated with him. As a result of his constant elusiveness, both within the film and real life, observation focuses on signs of his presence based on predictive profiling rather than monitoring of his actual physical being. To an extent, this involves analysing patterns of behaviour as proposed by Didier Bigo (2011). Engaging with the work of Bigo (2011), Foucault (1991), Mathiesen (1997), and Lyon (2003, 2011), as well as military surveillance theorists, this essay considers the multiple scopic regimes evident within the film and the tensions between them in respect of Foucault and Mathiesen. It pays particular attention to the relationship between surveillance, terrorism, and space in the contexts of post-9/11 America and argues that while surveillance in the film appears to be synoptic, at times, it synthesises the ideas of Foucault and Mathieson, and effectively constitutes a variant of these that intersects with Bigo's model.

THEORISING SURVEILLANCE

In contrast to Foucauldian panopticonism, Mathiesen (1997) describes a synoptic model which he considers to be inversely equivalent to panopticonism, stating that "as a striking parallel to the panoptical process, and concurring in detail with its historical development, we have seen the development of a unique and extensive system enabling *the many to see and contemplate the few*, so that the tendency for the few to see and supervise the many is contextualized by a highly significant counterpart [original emphasis]" (Mathiesen 1997, p. 219). He contends that such mass viewing typically occurs in relation to television and terms the outcome of this concurrent two-way panoptic/synoptic system of observation as a "viewer society" (Mathiesen 1997, p. 219). Akin to panopticonism, Mathiesen suggests that power and control are likewise features of synoptic viewing and argues that "in synoptic space, particular news reporters, more or less brilliant media personalities and commentators who are continuously visible and seen are of particular importance [...] They actively filter and shape information [...] they produce news [...] they place topics on the agenda and avoid placing topics on the agenda" (Mathiesen 1997, p. 226). In other words, these personalities exert power over viewers. So too does the media wield control and Mathiesen states that "synopticism, through the modern mass media in general and television in particular, first of all directs and controls or disciplines our *consciousness* [original emphasis]" (Mathiesen 1997, p. 230). Between the theories of Foucault and Mathiesen lies the concept of the 'ban-opticon', proposed by Didier Bigo which "indicate[s] how profiling technologies are used to determine who is placed under specific surveillance" (Bauman in Bauman and Lyon 2013, p. 61). This is pertinent here since the behaviour of the inhabitants of a residence associated with bin Laden is unusual and alerts suspicion, thereby leading to satellite observation and therefore extends Mathiesen's theory of synopticism.

For Aaron Doyle (2011), however, Mathiesen's parallel and reciprocal system of control and surveillance is problematic. Doyle argues that reducing surveillance to a state of observation is simplistic and does not take into account its multiple dimensions, such as resistance, which is an aspect that became central to later Foucauldian analysis and other studies of prison behaviour (Matthews 1999, p. 72). Following on from Doyle's comments concerning resistance, Lyon proposes that the more intense the panoptic surveillance, the greater the opposition to it, and refers to a "supermax" prison facility by way of example (Lyon 2011, p. 5). Lyon goes on to explain that "the more stringent and rigorous the panoptic regime, the more it generates active resistance, whereas the more soft and subtle the panoptic strategies, the more it produces the desired docile bodies" (Lyon 2011, p. 4). Such resistance can take multiple forms, which might extend from the creation of small pockets of private space, self-harming, and suicide to the reclamation of more significant areas of space (as in prison riots). Indeed, profound acts of resistance and chaos produced by acute repression through intimate surveillance and spatial confinement contribute to a balance of power such that "power is not a force exerted by a single individual but exists in equilibrium within a group" (Pheasant-Kelly 2013, p. 102). If this is evident in real-life carceral scenarios, it is often exaggerated in fictional circumstances for dramatic effect. Irrespective of its basis in real-life events, this is precisely the case in *Zero Dark Thirty* whereby the CIA's exertion of power through enhanced interrogation techniques leaves the first detainee that the spectator encounters, Ammar al-Baluchi (Reda Kateb), in a disorientated condition, and resistant to questioning.

In further discussion concerning resistance, Doyle also states that "surveillance has become much more widespread, dispersed far beyond the state, fragmented and multifarious" and does not necessarily involve direct viewing (Doyle 2011, p. 289). Indeed, the development of various technologies has led to scrutiny by means other than visual inspection. One mode of this monitoring involves the use of metrics, such as voice recognition and fingerprint analysis, aspects which are discussed elsewhere in this volume. In addition, bodily fluids afford a physiological mode of scrutiny that is not only predictive of certain physical characteristics, but is also able to pin down identity through DNA testing. In particular, developments in digital surveillance have had a profound effect on intelligence gathering during warfare, taking the form of robotics and unmanned aerial vehicles (Singer 2009). This is relevant in relation to *Zero Dark Thirty* since the progress of the final mission to capture bin Laden is monitored by satellite observation and its success rests on stealth- and night-time surveillance technologies. As well as these military applications, the expanded field of digital mapping in the real world extends from telephone taps through to deep-sea sonar scans and hospital monitors while electronic tagging of prisoners has facilitated a move from a disciplinary society to what Gilles Deleuze terms "a control society" (Deleuze 1995, p. 174). Drone technology has specifically promoted invisible aerial surveillance and, as Zygmunt Bauman notes, "Since 9/11, the number of hours needed by Air Force employees in order to recycle the intelligence supplied by the drones went up by 3100%—and each day 1500 more hours of videos are added" (in Bauman and Lyon 2013, p. 21). In relation to *Zero Dark Thirty*, telephone tapping is a key tactic utilised in the detection of bin Laden's courier and enables the team to locate broadly defined areas where he operates. The real life intensification of such surveillance resulted from the post-9/11 Patriot Act, which not only legitimised intrusions into privacy, such as the collection of phone records and Internet search data (Andrejevic 2007, p. 258) but also, as Andrejevic further notes "exempts monitoring activities under its aegis from the Federal Freedom of Information Act" (Andrejevic 2007, p. 7).

SURVEILLANCE AND *ZERO DARK THIRTY*

Zero Dark Thirty charts the search for bin Laden after the September 11 attacks in New York and follows the quest of a military operative, known only as Maya (Jessica Chastain), to locate him. Scholarly studies of the film tend to focus either on its interrogation methods, which invoked controversy for implying that torture was effective in tracing bin Laden (McSweeney 2014; Tanguay 2015), or its characterisation, specifically that of Maya (McSweeney 2014). Otherwise, there is limited analysis of surveillance methods within the film. As outlined above, the film's modes of monitoring take multiple forms: namely, the observation of the enhanced questioning of detainees, directly by interrogators and indirectly, by those viewing on CCTV; ground level physical surveillance of the compound's exterior and surrounding areas of Pakistan where bin Laden is thought to be located; photography of suspects; mobile phone interception and location tracking via digital methods; infra-red surveillance night vision goggles; satellite and drone observation of the compound; and predictive patterning of behaviour. If Foucault's panoptic concept depends on a fixed central point of observation and presence at a certain location for being observed, and Mathiesen's model of synoptic observation hinges on the observed being located at a specific place in space and time (albeit mediatised space), in a sense, neither model is entirely appropriate to *Zero Dark Thirty*. This is because its many points of surveillance and sites of observation are constantly mobile and mutable. Moreover, there is a persistent oscillation between who is observing and who is being observed as instances of terrorist resistance arise.

The blank screen opens with a series of air-traffic controller recorded reports concerning the hijacked planes on September 11 2001, and real-world telephone conversations from those trapped inside the World Trade Center and those aboard the hijacked planes. Even as the film denies the viewer visual access to the events that escalated the search for bin Laden, the recordings provide an example of auditory surveillance in their monitoring of the flight paths of the hijacked planes and the activities of individuals aboard, as well as the suffering of those involved in the Twin Towers. The immediately ensuing two-year flash-forward to a sequence entailing the torture of the aforementioned detainee, Ammar al-Baluchi, leads one to assume that he is associated with 9/11. For Liane Tanguay, such an immediate juxtaposition "of the attacks with the brutal 'enhanced interrogation' sequence that follows cannot but condone what America repeatedly insisted it 'did not do', namely torture" (Tanguay 2015, p. 300). In a related vein, David Jones and Mike Smith argue that the film exists on a broader continuum of 'dark' films made after 9/11 that they label 'Dark Americana', which "stands in contrast to a dominant mode of cinematic portrayal that asserts largely positive American self-imagery [...]. Dark Americana recognises a world of moral ambiguity and emotional complexity. Here, after 9/11, we see the evolution of an artistic current that perceives political life existing in degrees of dysfunction" (Jones and Smith 2016, p. 3). The scenes of 'enhanced interrogation' take place in a CIA prison at an, as yet, undisclosed black site where al-Baluchi is first questioned by CIA interrogator, Dan (Jason Clarke) before being strung up with his arms outstretched in the now familiar stress position forced upon detainees that came to public attention in 2004. "I own you. Look at me!.....Look at me!" shouts Dan, as if the act of coercing al-Baluchi to gaze directly at him will render him more susceptible to questioning. Another operative, clad in a black balaclava and protective over-garment, and whose identity so far remains unknown, stands by and watches the interrogation. Thereafter, the spectator, along with various other CIA members, observes the still-chained detainee via a CCTV screen outside the interrogation room before the balaclava clad operative peels off head-wear and protective gear to reveal her identity. Maya, who is a newly trained CIA intelligence analyst, becomes the film's central protagonist, her single-minded determination to find bin Laden setting the course of events. As the group of CIA operatives stands outside the interrogation room, we see Maya glance again at the screen, which is disclosed via her subjective viewpoint. Given her prior apparent concern about the torture of al-Baluchi, one might interpret her anxious look at the monitor as one of sympathy, although she confounds such expectations by indicating to Dan that they should go back into the interrogation room. Dan too registers her seeming unease about the procedures, by stating "there's no shame if you want to watch from the monitor". However, Maya not only returns to the interrogation room, but also refuses to wear the black balaclava, the film therefore implying that she is tough and ruthless. Yet, as the camera continues to intercut between the ongoing torture of al-Baluchi and the observing Maya, her figure behaviour suggests otherwise. Framed in close-up, we see her swallow, as if anxious, and she repeatedly averts her gaze and visibly winces as Dan 'water-boards' al-Baluchi by deploying a highly controversial interrogation technique that was commonplace at CIA black sites and came to be considered as torture (Danner 2004). According to Terence McSweeney, who is critical of the way that torture is portrayed in the film, Maya represents "the face of the West" (McSweeney 2014, p. 45). Certainly, the visual scrutiny of her face through frequent close-ups invites viewer identification with a character who presents a combination of vulnerability (in her stature, fragile features, and the way she is later attacked by al-Qaeda), resilience, and fortitude. As McSweeney further comments "the film would have us believe that Maya is the film's true victim and in this way she may be the quintessential symbol of America's understanding of the war on terror" (McSweeney 2014, p. 44). Even so, shortly thereafter, we learn from CIA head, Joseph Bradley (Kyle Chandler), that "Washington says she's a killer", suggesting how readily one misinterprets images and exercises an unconscious bias.

The CCTV that monitors al-Baluchi is in an elevated position within the interrogation room and therefore views him from a high angle perspective. This inherently promotes the notion of a watchful gaze over a subject who does not know when he is being observed and therefore has features akin to the central tower of the panopticon. However, in this case, the fact that several operatives scrutinise one individual on a screen concurrently conforms to Mathiesen's model. Strategies of interrogation subsequently involve the removal of his clothes—to promote humiliation—and sleep deprivation, which lead not only to resistance, but also to incoherency, and when Dan leaves the room, al-Baluchi appeals to Maya to help him. Robert Burgoyne detects a momentary tension between the two, which, for him, is articulated via Maya's gaze rather than touch. As Burgoyne notes, "It is her gaze that is invoked as the prisoner is paraded naked from the waist down and then walked round the room like a dog; it is her gaze that holds the power to humiliate, and for which the scene is staged; and it is her gaze that Dan [...] consults before he continues the torture" (Burgoyne 2014, p. 249). In response to his pleas and seeming indifferent to his state of undress, Maya calmly approaches the detainee and coldly tells him that he can help himself by being truthful. Subsequently, Maya and Dan trick al-Baluchi, who becomes increasingly incoherent because of sleep deprivation, into thinking that he has confessed. For some critics and scholars, this sequence suggests that torture was successful in leading the CIA to bin Laden and has attracted controversy because of its deviation from reported actual events. McSweeney highlights other shortcomings, specifically, the stereotyped representation of protestors outside the Embassy of the United States in Pakistan as "a violent, threatening horde", while "every single member of the ethnically and gender diverse CIA are shown to be industrious, ethical and patriotic" (McSweeney 2014, pp. 38–39). In a related vein, Guy Westwell contends that "Once again, an example of post-9/11 cinema indicates how seemingly irreconcilable views of 9/11 are drawn together in alignment with a hegemonic historical revision that, while acknowledging some ambiguity, restores credibility for US national identity as a whole" (Westwell 2014, p. 178).

From then on, Maya views videotaped interrogation interviews between Dan and other suspects. We are reminded of the surveillant aspect of her work by the low resolution of the visuals and the sound quality, which is slightly distorted. The camera cuts between these video quality images, which mostly maintain a high angle perspective (thereby reflecting the position of the CCTV but also continuing to mirror a panoptic arrangement), and extreme close-ups of Maya as she scrutinises them, one assumes, for clues that will lead to bin Laden. At times, the frame of the diegetic screen corresponds to the cinematic frame, with Maya in the foreground viewed from the rear as she watches the filmed torture. Arguably, the sequence enables not only Maya but also the spectator to linger over the torture. In short, she becomes a surrogate for the spectator, implicating the viewer in the events. Such sequences raise certain issues in respect of surveillance: foremost, they magnify the ethical implications of viewing torture—for, just as Maya as a viewer is complicit in the suffering, so too is the extra-diegetic spectator incriminated, effectively for watching torture as a leisure activity. In this respect, Jay Bernstein asks whether "is not the photograph of the body in pain [...] a further exploitation of it, a repeating of the injury done to the victim which does nothing for her while providing the viewer with the pleasures of affective intensity without the cost of ethical responsibility?" (Bernstein 2012, p. xii). While Bernstein is referring to images of real atrocity, there is nonetheless a line of continuity between the historical torture enacted at the terrorist detainee camps, and the representation of that torture in a film that presents an allegedly authentic version of real events. Although sexual abuse of the prisoners—as documented at the detainee camps—is not represented in the film (this omission, in itself, suggesting a biased filmmaking perspective and supporting the claim of Westwell), one might posit that the repeated images of suffering and distress promote an eroticisation of the tortured body, though, similar to the Abu Ghraib photographs, the problem here is less the pleasure in looking and more "the moral indifference of the [images]" (Butler 2009, p. 91). In a related vein, the second issue concerns the unconscious discrimination and uncertainty associated with human intervention as opposed to the objectivity of computerised systems and biometric measurements, particularly significant given that the sequence entails a white American woman scrutinising the suffering of non-white male detainees. Notwithstanding the potentially racial and gendered elements inherent in this scenario, the pressure to locate bin Laden after 9/11 undoubtedly had a major emotional impact on objectivity in the real world situation, an aspect which constantly permeates the film. For instance, when meeting a potential informer, Humam al-Balawi (Musa Sattari), at Camp Chapman CIA facility in Afghanistan, Jessica (Jennifer Ehle), Maya's colleague, is so desperate to track bin Laden that she does not have him searched when he enters the Camp—however, al-Balawi is a suicide bomber and blows himself up, killing seven CIA operatives/soldiers, including Jessica. As Kevin Haggerty notes "it matters enormously who is actually conducting surveillance [...] The attitudes, predispositions, biases, prejudices and personal idiosyncrasies of the observers can be vitally important in shaping the form, intensity and regularity of those responses" (Haggerty 2011, p. 33). In addition, the inability to interpret imagery is often evident throughout the film. For instance, in the previously mentioned sequence, as Maya physically moves closer to the monitor as if to discern the details of what is being said (mediated through subjective viewpoint), the picture resolution deteriorates still further and the detainees' faces become indistinct so as to visually signify the absence of useful intelligence. A colleague's question shortly thereafter of "how's the needle in the haystack?" sums up the repeated difficulties in securing credible information.

As if to illustrate the failure to obtain actionable intelligence, the film cuts to a scene of the 2005 London bombing, and then between a Bagram black site and Pakistan where Maya herself now takes over the interrogation of various suspects. Here, the overriding inability to locate a prime suspect, a possible courier (Abu Ahmed [Tushaar Mehra]) who is thought to communicate between bin Laden and one of his contacts, Abu Faraj al-Libbi (Yoav Levi), indicates not only their abortive intelligence gathering, but also the success of al-Qaeda's resistance. In fact, one interviewee tells her "you will never find him—he is one of the disappeared ones". Whilst in Pakistan, Maya arranges to meet Jessica at the Islamabad Marriott hotel. Despite the detailed scrutiny that Maya undergoes when entering the hotel gates, a bomb explodes there, again indicating the fallibilities of human surveillance and the overall shortcomings of the CIA's enterprise.

Photographic evidence too is fundamentally flawed and proven unreliable (though this detail is absent from Bergen's account of the real events), indicated by the fact that Maya and a colleague come to realise that Abu Ahmed, the suspect they found impossible to trace, and who was then believed to be dead, had eight brothers who all looked alike. It transpires that their photograph is actually of one of Abu Ahmed's brothers and so they had been looking for the wrong person. This corroborates a claim for a lack of objectivity in human observation, since facial recognition technology might have correctly identified the individual. The fact that information concerning Abu Ahmed's existence only emerges by chance in 2007, a detail that is substantiated in Bergen's account (2013, p. 122), also suggests the shortcomings of the Total Information Awareness[1] project (both within the diegesis and in reality), which was established in 2002 following 9/11. Effectively, one might argue that Maya and her colleagues see what they want/need to see—a case readily borne out in real life by the misinterpretation of images of alleged weapons of mass destruction in Iraq. Indeed, as Lyon contends, technical modes of screening do not have the discriminatory aspects and uncertainty associated with human intervention and explains that "This is because system designers and computer programmers play a greater role in creating the categories, which are the criteria for discrimination. In other words, the processes by which unusual or abnormal behaviours are defined are tasks for "technical experts" rather than ones in which there is ethical scrutiny or democratic involvement" (Lyon 2003, p. 135).

In order to locate Abu Ahmed, Dan bribes a Kuwaiti contact for the telephone number of the suspect's mother. Consequently, the CIA tap the phone and we see close-ups of charts of sound recordings of their voices as they talk, conveying again a visual representation of auditory surveillance. The static, full screen image of the chart monitoring their conversation then cuts to rapid tracking shots that follow telephone wires as if we are literally tracing the conversation of the suspect through the connections, as another signifier of auditory surveillance. Thereafter, a cut to Google Earth imagery zooms into an area of Rawalpindi, thereby pinpointing the alleged location of Abu Ahmed. Street names are visible on screen, emphasising the increasingly sophisticated scope of digital 'watching'. Having identified two potential locations for Abu Ahmed, the team, headed by Maya, undertakes ground level physical surveillance based on telephone interception. Here, close-ups of their computer screen show the radius within which telephone calls are being made by Abu Ahmed ('magic box' surveillance). At times, surveillance involves the team walking on foot and physically looking for the suspect by visually scanning for those using a mobile telephone, or driving towards them wherever the mobile signal becomes stronger. In this ironic scenario, Maya and her team do not know who they are looking for—and could be unwittingly watching their suspect. A rapidly edited combination of high angle shots, and low-level street shots of the marketplace, and close-ups of the computer monitor cause a confusing sequence that conveys the difficulties of tracking a moving target whom the team cannot physically see. At one point, the driver, Larry (Édgar Ramírez) leaves the vehicle and scans the crowds for anyone using a mobile phone. A long shot reveals him surrounded by crowds before the camera cuts to an extreme long shot to emphasise the size of the throng and the daunting nature of their task. Because the team are amidst teeming streets and squares, one might argue that the few *do* watch the many, but not in the manner prescribed by Foucault. Rather, they look in a random, non-systematic manner based on chance, and, founded on the possibility that they *may* be looking at their suspect, they adjust their actions accordingly, creating, in a sense, an inverse relationship with the prisoner in the Foucauldian panopticon. Moreover, when Jack (Harold Perrineau), one of the CIA analysts, maps all the locations from which the suspect makes calls, he is unable to identify a pattern to the site and duration of the calls. Even as they are trying to use predictive surveillance to find patterns that will enable them to ascertain Abu Ahmed's next move, his movements are markedly different to a Foucauldian model wherein activities are coordinated, precise and occur according to predictable schedules. It is only when the CIA team fix their own point and remain stationary, and instead wait for increases in the suspect's telephone signal strength that they succeed in locating the courier (who is driving a distinctive white SUV). By physically watching who is entering the near vicinity, they manage to photograph him. The next line of surveillance involves using locals positioned alongside a road to track the vehicle's movements. We are made aware of their surveillant roles because camera close-ups that cut to extreme close-ups track shifts in their eye-line and head movements as they watch the white SUV and then, as soon as it has passed by, they either make written notes or phone in their locations.

Having determined the residence associated with the SUV, the scene cuts to monitor low resolution satellite images of the compound. The compound, revealed to be at Abbottabad, is now visible remotely on three large-scale screens in an operations room at Predator Bay, the CIA Headquarters at Langley. As the CIA officials view the indistinct individuals moving around within the compound, the females are distinguished by the fact that they hang laundry on a clothes line (deduced because "men don't mess with the wash"), and children, by virtue of their size relative to animals grazing in the adjacent area. Having witnessed three females and only two males, the CIA concludes that there is a third male living there, whom they believe might be bin Laden. But, despite multiplatform surveillance that even goes to the extent of trying to obtain biological evidence, they are unable to ever see him or visualise him, merely recording him as a "heat signature". This is because he exercises extreme measures to effectively resist detection, by destroying any proof of his bodily presence. For example, we learn that the garbage from the compound is burnt so that the CIA is unable to get a DNA sample; there are no telephone calls or Internet connections made from the compound; and he does not ever leave the compound. The section chief, George (Mark Strong), states that they even considered going into the sewer to obtain a faecal sample for DNA evidence as a means to deduce his identity. As Lyon states "In numerous surveillance situations, bodies are reduced to data [...] One cannot but conclude that information *about* that body is being treated as if it were conclusive in determining the *identity* of the person [original emphasis]" (Lyon in Bauman and Lyon 2013, p. 134). While Lyon refers to biometric means of identification at border crossings, there is nonetheless a continuity with the means of surveillance exercised in the context of bin Laden. In addition, there is arguably a similar element of dehumanisation. As Lyon continues, "Ironically, though, surveillance—someone to watch over me— may well be valued and sought after in the vicissitudes of liquid modern life. Unfortunately [...], however, that 'someone' is all too often some- *thing*. And the something is supposedly disembodied information, sorting by means of software and statistical technique. It's the product of double adiaphorization, such that not only is responsibility removed from the process of categorising, but the very concept of information itself reduces the humanity of the categorized, whether the end in view is dating or killing [original emphasis]" (Lyon in Bauman and Lyon 2013, p. 137). In a similar vein, the heat signatures that stand in for the inhabitants of the compound at Abbottabad also reduce their humanity. While the heat signatures are identifiable according to their gender roles, or size in com- parison to animals, the Navy Seals' 'real-time' encounter with them during the raid on the compound evokes a very different effect. As one Seal enters the compound following the killing of bin Laden, a close up discloses his troubled facial expression followed by subjective viewpoint, which reveals bleeding corpses and crying children. This brief moment of "affective experience and somatic intensity" (Burgoyne 2014, p. 247) exhibited by the Seal gives insight into the inhabitants as 'real' people and as a family, and is distinct from the aerial impressions created previously.

Clearly therefore, surveillance does not always guarantee identification, a pattern implied throughout the film in the form of mistaken and obscured identities and, overall, there is a tension between watching and being watched. If, as one CIA senior supervisor tells his chief, "This is a pro- fessional attempt to avoid detection," so too does the CIA take measures to evade identification. Maya herself wears a disguise to conceal her identity; regardless, she is obviously being watched, because she is attacked one morning by gunmen as she leaves her residence. Likewise, in the attack on bin Laden's compound, the Navy Seals use modified Black Hawk stealth helicopters, which have decibel mufflers on their blades, and execute their mission under cover of night. Throughout the operation, Maya watches from a monitor as they cross into Pakistan from Jalalabad in Afghanistan. Extreme overhead shots track the trajectory of the helicopters, which fly quietly in conditions of low light although after the aircraft come into 'sight' on the screen, one crash lands because of unanticipated down-drafts created by the height of the surrounding walls. When the Seals enter the compound, the spectator assumes their perspective via night vision goggles through which we see grainy, low-resolution, green-toned imagery and therefore, similar to other modes of surveillance deployed in the film, the spectator is positioned as an observer from a US standpoint. If this affords the spectator insight into the difficulties of tracking bin Laden, it also endows us with the same lack of knowledge as the CIA. Though as a real event we know the outcome, the film specifically informs us of the absence of conclusive proof that bin Laden was living in the compound. Indeed, following Foucauldian concepts, surveillance in itself does not lead them to bin Laden. Rather, it is knowledge of cultural traits ("Muslim women either live with their parents or with their husband"), bin Laden's relationships, and the typical tendencies of high-profile terrorists (known as 'tradecraft') that enable his capture. A significant aspect of his detection also depends on Maya's instincts that he is '100%' at the compound, which generally contradicts accepted scholarship concerning human intervention, although there are reports of subjective surveillance through "personal interrogation and assessment of passengers" being highly effective in real-world Israeli airport monitoring (Lyon 2003, p. 134). Maya's input is not a feature of Bergen's book on the search for bin Laden, but he does include a comment by Michael Scheuer, the founder of the bin Laden unit, on the significant contribution of female CIA operatives that stated "They seem to have an exceptional knack for detail, for seeing patterns and understanding relationships" (Bergen 2013, p. 77).

In these observations concerning the relationship between bin Laden, space and identity, one might argue that the compound effectively functions as a prison, one distinction obviously being the fact that he is not observed by a prison guard from a central tower. Even so, the constraints on his movement and his restriction to the compound arise as a result of criminal transgression. Moreover, though he is not scrutinized from within the compound, he is constantly monitored by the satellite, and the high walls and bolted doors, as well as his prescribed movements, resemble those of prison life. Also similar to the Foucauldian panopticon is the elevated single point of surveillance, the satellite being located directly over the compound, albeit this single point of observation is dispersed to multiple viewers stationed in a distant CIA operations room. Because his social interaction and social space is limited, one might also argue that, in a similar manner to prison inmates, bin Laden's sense of identity may be compromised. In short, he must police his movements to avoid observers that may be watching, even though remotely, from a distance geographically afforded by the drone. Bergen makes a similar analogy regarding the real-world scenario and states that "The pacer [bin Laden] never left the compound and his daily excursions seemed like those of someone in a jail yard who couldn't leave but was trying to get some exercise. He walked very rapidly in tight circles, then went back inside" (Bergen 2013, p. 132). Arguably, this circumstance synthesizes the ideas of Foucault and Mathieson in the way that bin Laden knows he may be being watched, but does not know when; he is watched by observers removed geographically (akin to Mathiesen's concept of television viewing); and he is constrained to a high-walled, impregnable fortress. Accordingly, the film's deployment of surveillance impacts upon spectator understandings of the relationship between geographical and bodily spaces—the compound is viewed constantly in distant CIA headquarters in Langley, United States, and this continuity means that those spaces occupied by the body are extremely limited. Even in death, bin Laden's body is symbolically constrained, physically by concealment from the spectator/his family in a body bag, and visually by acute camera angles. This is partly to sustain the film's verisimilitude and credibility, but also, as previously noted, reflects the way that images of his body were not available on media at that time.

Therefore, whilst having commonalities with the surveillant models proposed by Mathiesen and Foucault, and arguably being a synthesis of them, the monitoring of bin Laden and other al-Quaeda suspects is modified by recent technologies. These no longer depend on direct viewing, and mean that remote access and "the 'comfortable invisibility' of these eyes in the skies" and the "indirect effect that those countries or states using such at-a-distance technologies thereby also distance themselves from the conflicts, crimes or crises they are supposed to detect or deter" (Lyon in Bauman and Lyon 2013, p. 77) are distinct from Foucauldian notions of panopticism. Lyon also detects a subtler difference between traditional Foucauldian and new surveillance in terms of feelings of security. According to him, "The prominent means of procuring security, it seems, are new surveillance techniques and technologies, which are supposed to guard us, not against distinct dangers, but against rather more shadowy and shapeless risks. Things have changed, for both watchers and the watched. If once you could sleep easy knowing that the night watch was at the city gate, the same cannot be said of today's "security". It seems that, ironically, today's security generates forms of *in*security as a by-product [original emphasis]" (Lyon in Bauman and Lyon 2013, p. 101). Related to this, Bigo suggests that security has led to "the field of "unease management", which is the formation of global police networks, policing military functions of combat and criminalizing the notion of war. The governmentality of unease is characterized by practices of exceptionalism, acts of profiling and containing foreigners, and a normative imperative of mobility" (Bigo 2011, p. 47).

CONCLUSION

In summary, the film reveals a number of surveillance strategies, ranging from conventional forms, such as direct observation during interrogation, physical pursuit, and CCTV monitoring, to more sophisticated technologies, including stealth helicopters, night vision goggles, drones and mobile phone tracking. This chapter has examined the ways in which filmic representations of such methods reflect changes in real-world monitoring of both public and terrorist activities since 9/11. Indeed, the film epitomises the foregrounding of surveillance in visual culture since the New York attacks. It also gives insight into a historically significant event, especially the difficulties and moral issues in the operation to locate bin Laden. If the capture of bin Laden was determined by human tenacity, many other situations involving human input led to errors that proved to have serious or fatal consequences, supporting a claim for the superior objectivity of new surveillance technologies. One might further argue that the revelations of the film highlight ethical questions concerning the proliferation of surveillant methods such as phone tapping. If the film is open to accusations of bias in representation, one might also question the display of torture for spectator pleasure. In evaluating the film's modes of surveillance, there are intersections between both Foucault's and Mathieson's models whereas the final focus on the compound itself may be explained with reference to Bigo's work. However, as articulated in Foucauldian approaches, there is a balance of power between those who watch and those who are watched. Power exerted by the observer causes resistance in those observed and in the case of bin Laden, this entails the complete evasion of detection. Construction of the compound where he lives, itself spatially designed to facilitate this, enables such resistance whilst also allowing him to orchestrate the al-Qaeda network via a courier. Its walls operate much like the prison, but rather than keeping people in, it aims to keep others out. The CIA and the terrorist detainee camps also operate on panoptic principles but deploy CCTV in lieu of a central tower. However, neither the black sites nor bin Laden's compound reflect the orderliness associated with Foucauldian concepts of the institution, but rather, become transformed into sites of transgression and abjection. At the detainee camps this occurs in attempts to dehumanise inmates through enhanced interrogation techniques that involve hooding, stress positions, water-boarding and shackling. While previous studies often focus on these controversial aspects, analysis of surveillance in the film highlights necessary aspects of contemporary monitoring. Despite the failures of surveillance present in the film (and in real-life), and the fact that the CIA was not completely sure that bin Laden was living in the compound, their mission was successful. As Bigo suggests, September 11 has legitimised surveillance and "has reinforced the idea that the struggle against these threats justifies the profiling of certain people's potential behaviours, especially if they are 'on the move'. The political reaction to September 11 justifies a proactive and pre-emptive strategy, which has the ambition to know and to monitor the 'future'" (Bigo 2011, p. 63). Overall, the film gives insight into surveillance as essential to the war on terror. Yet, ultimately, the mission to kill bin Laden was equally founded on deduction and profiling, with its success, in both real and fictional scenarios, being one of probability rather than absolute guarantee.

NOTE

1. David Lyon describes how the Department of Homeland Security established a comprehensive counter-terrorism database with the aim of detecting, classifying and identifying foreign terrorists, which is termed as Total Information Awareness (2003, p. 91).

REFERENCES

Andrejevic, M. (2007). *iSpy: Surveillance and power in the interactive age*. Kansas: University of Kansas Press.
Bauman, Z., & Lyon, D. (2013). *Liquid surveillance*. Cambridge: Polity Press.
Bergen, P. (2013). *Manhunt: From 9/11 to Abbottabad—The ten-year search for Osama Bin Laden*. London: Vintage Books.
Bernstein, J. (2012). Preface. In Asbjørn Gronstad & Henrik Gustafsson (Eds.), *Ethics and images of pain* (pp. xi–xiv). London: Routledge.
Bigo, D. (2011). Security, exception, ban and surveillance. In David Lyons (Ed.), *Theorizing surveillance: The panopticon and beyond*. London: Routledge.
Burgoyne, R. (2014). The violated body: Affective experience and somatic intensity in zero dark thirty. In David LaRocca (Ed.), *The philosophy of war films* (pp. 247–260). Lexington: University of Kentucky Press.
Butler, J. (2009). *Frames of war: When is life grievable?* London: Verso.

Danner, M. (2004). *Torture and truth: America Abu Ghraib and the war on terror*. New York: New York Review Books.

Deleuze, G. (1995). *Negotiations*. New York: Columbia University Press.

Doyle, A. (2011). Revisiting the synopticon: Reconsidering Mathiesen's "The viewer society" in the age of web 2.0. *Theoretical Criminology, 15*(3), 283–299.

Foucault, M. (1991). *Discipline and punish: The birth of the prison*. London: Penguin.

Haggerty, K. (2011). Tear down the walls: On demolishing the panopticon. In David Lyon (Ed.), *Theorising surveillance: The panopticon and beyond* (pp. 23–45). London: Routledge.

Jones, D., & Smith, M. (2016). The rise of Dark Americana: Depicting the "War on Terror" on-screen. *Studies in Conflict and Terrorism, 39*(1), 1–21.

Lyon, D. (2003). *Surveillance after September 11*. Cambridge: Polity Press.

Lyon, D. (Ed.). (2011). *Theorising surveillance: The panopticon and beyond*. London: Routledge.

Mathiesen, T. (1997). The viewer society: Michel Foucault's panopticon revisited. *Theoretical Criminology, 1*(2), 215–234.

Matthews, R. (1999). *Doing time: An introduction to the sociology of imprisonment*. London: Macmillan.

McSweeney, T. (2014). *The 'War on terror' and American film: 9/11 frames per second*. Edinburgh: Edinburgh University Press.

Nolan, C. (2008). *The Dark Knight*. USA/UK.

Pheasant-Kelly, F. (2013). *Abject spaces in American cinema: Institutional settings identity and psychoanalysis in film*. London: IB Tauris.

Singer, P. W. (2009). *Wired for war: The robotics revolution and conflict in the 21st century*. London: Penguin.

Spielberg, S. (2002). *Minority Report*. USA.

Tanguay, L. (2015). The "Good war" on terror: Rewriting empire from George W. Bush to American sniper. *Critical Studies on Security, 3*(3), 297–302.

Westwell, G. (2014). *Parallel lines: Post-9/11 American cinema*. London: Wallflower Press.

AUTHOR BIOGRAPHY

Frances Pheasant-Kelly is Reader of Film and Television Studies at Wolverhampton University. She is the MA course leader in Film Studies, and director of the Centre for Film, Media, Discourse and Culture. Her research focuses on 9/11, space and cinema as published in a co-edited collection entitled *Spaces of the Cinematic Home: Behind the Screen Door* (Routledge, 2015) and two monographs which include *Abject Spaces in American Cinema: Institutions, Identity and Psychoanalysis in Film* (Tauris, 2013) and *Fantasy Film Post 9/11* (Palgrave, 2013).

CHAPTER 6

To See and to Be Seen: Surveillance, the Vampiric Lens and the Undead Subject

Simon Bacon

SEEING AND BEING SEEN

Dracula, whilst ostensibly about the invasion of Victorian society by a vampire from Transylvania, in fact exemplifies the ways in which surveillance was used to observe and control societal identity both in and outside of the British Empire. A large proportion of the action takes place in England, where Professor Van Helsing takes charge of the team arrayed against the vampire, with many devices employed to observe and record the main players during the tale. As noted by Leblanc, Van Helsing takes "strict control of the dispersal of vampiric discourse" (Leblanc 1997, p. 256), which involves him reading all the material produced by the rest of the group. Mina Harker, in particular, collates this information and, as Ken Gelder observes, rather than she controlling it, "her own writings are overseen by Van Helsing: women (and indeed men) are under surveillance in this novel" (Gelder 1994, p. 81). This surveillance is facilitated not just by Mina's secretarial skills, but the many forms of technology used in the

S. Bacon (✉)
Poznan, Poland
e-mail: baconetti@googlemail.com

© The Author(s) 2017
S. Flynn and A. Mackay (eds.), *Spaces of Surveillance*,
DOI 10.1007/978-3-319-49085-4_6

novel: Telegram, photograph and phonograph to name but a few. All these recording Body/Tel, or observational machines, form a kind of undead gaze in being entities that are neither alive or dead, but are active participants in the narrative, not unlike the vampire itself. Observation here not only controls information, but the identity of the group as a whole as well, or, as Leblanc further comments, discourse, and indeed all mechanism of observing and recording, are "subject to social control" (Leblanc 1997, p. 256). If Van Helsing exemplifies close observation and control, then Count Dracula's influence over his homeland shows how a culture of surveillance embodies extremes of the all-seeing-eye and the panopticon in ways that change those that know they are being watched; the effects on the inmate of continual surveillance instilling "in the labouring classes a distinctive temporal and bodily discipline" (Lyon 2006, p. 27).

The vampire's influence upon the people around the vicinity of its home causes exactly the same results as those described by Lyon, changing everyone's pattern of behaviour and even the nature of their identities. The sovereign gaze of Count Dracula spreads far and wide in Transylvania and, whilst it is seen in the figure of the vampire, it is mainly manifested in the "eyes" of its castle lair that is situated like an observation tower where it can view as much of the land around it as possible, as described in Stoker's book:

> Then we looked back and saw where the clear line of Dracula's castle cut the sky…We saw it in all its grandeur, perched a thousand feet on the summit of a sheer precipice, and with seemingly a great gap between it and the steep of the adjacent mountain on any side. There was something wild and uncanny about the place (Stoker 1996, p. 404).

The blank stare of the vampiric watchtower is observed by the castle's newest arrival, Jonathan Harker, the young solicitor who unknowingly will facilitate the Count's invitation into London. "The driver was in the act of pulling up the horses in the courtyard of a vast ruined castle, from whose tall black windows came no ray of light, and whose broken battlements showed a jagged line against the moonlit sky" (Stoker 1996, p. 15) The gaze of the vampire emitted from these "black windows" affects the population for miles around. When Harker arrives at his hotel in Bistritz, which is some distance away from his final destination, he receives an unexpected response: "When I asked him if he knew Count Dracula, and could tell me anything of his castle, both he and his wife crossed themselves, and, saying

that they knew nothing at all, simply refused to speak further"
(Stoker 1996, p. 4) The anxiety caused by the mentioning of the Count
and/or his castle increases the closer the young solicitor gets to the
Carpathian Mountains and Castle Dracula, as shown in the exchange
below, from Browning's film, between him and an innkeeper at his last stop
before the last leg of his journey—in the film, the character Renfield
replaces Harker as the one travelling to the vampire's lair:

> Innkeeper: We people of the mountains believe at the castle there are vam-
> pires. Dracula and his wives, they take the form of wolves and bats. They
> leave their coffins at night and they feed on the blood of the living.
>
> Renfield: Oh, but that's all superstition. Why, I can't understand why...
>
> Innkeeper: Look. The sun. When it is gone they leave their coffins. Come.
> We must go indoors. (Browning 1931)

This also describes the ways in which the gaze of the vampire affects the
behaviour of those being observed by it, and also those that do not follow
the "rules" and, as punishment, will become victims of the undead Count
and his three wives. The inhabitants of Castle Dracula then enact punishment on behalf of the "remote" eyes of the watchtower, looking for
those that have not obeyed the sovereign law of the vampiric gaze.

Within the narrative then, the gaze of Dracula's castle is used as an
extension of that possessed by the vampire King himself. The importance
of this controlling gaze in Browning's film cannot be over emphasized and
consequently the film goes to great lengths make a feature of Count
Dracula's—and Bela Lugosi's, the actor who played him—eyes. As noted
by Gary D. Rhodes, "The 1931 film *Dracula* even used spot lighting to
highlight his eyes and eye makeup. Lugosi's performance brought Bram
Stoker's accent on Dracula's gaze in the (1897) novel into a powerful
cultural image"[1] (Rhodes 2006, p. 30) The Count's eyes, which quite
literally shine onscreen, capture and hold all those they focus on within
their hypnotic gaze. The persistence and power of the vampiric gaze is
described by Stacey Abbott as follows:

> It is a common characteristic of the vampire to stalk his or her victims, as
> demonstrated by the sequence in Stoker's novel wherein Mina and Jonathan
> [Harker] witness Dracula following the young woman in Piccadilly Circus,

his eyes fixed upon her every move. In this sequence, Stoker evokes a broader nineteenth-century understanding of the power of the gaze. In his discussion of superstitions about the evil eye and their relationship to George Du Maurier's *Trilby*, Daniel Pick points out that fear of the evil eye was still prevalent in the nineteenth century. (Pick 2000, pp. 178–179) The eyes were believed to have a physical effect upon the observed, possessing the power to paralyze, damage, or bring about bad luck...this gaze is equally present within Stoker's Dracula and is visually presented in the filmic vampire genre. (Abbott 2007, p. 94)

In this way the "eye" of the vampire not only embodies the panopticon, but also becomes a very real actor in the shaping of one's behaviour, and indeed identity, as not only does it make sure that the subject feels constantly under observation but is also able to administer punishment to those seen to not be adhering to sovereign, social conditioning. The most obvious example of this change of identity under the vampiric gaze in the novel is Harker himself. He has a very different character at the start of the story, as noted by Richard Astle: "while Jonathan waits to be admitted to Dracula's Transylvanian castle in the first chapter, he thinks of the fact that 'just before leaving London I got word that my examination was successful and I am now a full-blown solicitor!'...[exampling] Harker's assumption of a position in society" (Astle 1979, p. 99). However, once he has become a victim of the vampiric gaze, he is a totally different person, as related by a Sister at the hospital in Budapest where Jonathan is recuperating after escaping the surveillance of Dracula:

> He has had some fearful shock—so says our doctor—and in his delirium his ravings have been dreadful; of wolves and poison and blood; of ghosts and demons; and I fear to say of what. Be careful with him always that there may be nothing to excite him of this kind for a long time to come; the traces of such an illness as his do not lightly die away. (Stoker 1996, p.107)

No longer the confident young solicitor, his encounter with the vampiric gaze has dramatically changed him, as described by Mina:

> I found my dear one, oh, so thin and pale and weak-looking. All the resolution has gone out of his dear eyes, and that quiet dignity which I told you was in his face has vanished. He is only a wreck of himself, and he does not remember anything that has happened to him for a long time past. (Stoker 1996, p. 112)

In fact, Harker was almost turned into a vampire, a fate which is even further seen in Browning's film where Renfield—the Harker replacement in the story—becomes zoophageous and begins to eat insects. Further, Dracula's observation of Renfield in the film forms a telepathic bond between them so that the Count is aware of where he is and what he is doing even when hundreds of miles away. This begins to foreshadow conceptions of surveillance in the twenty-first century where the subject becomes self-surveilling and even when the lens/gaze of observation is no longer present, the newly created identity of the observed remains in place. Violence, and observable punishment, or what Foucault calls spectacle, are crucial in the enforcement and sustaining of sovereign law, as observed by Lorna Hutson, "the political spectacle of intense pain that reaffirms the body of the king and hence the body politic" (Hurston 2005, p. 33). In Stoker's novel and Browning's film, the most visible punishment Dracula can enact is turning the subject into a vampire, as seen most graphically in the character of Lucy Westenra from Stoker's novel, who becomes one of the undead, and in the almost vampirism of Jonathan Harker and/or Renfield.

Overall *Dracula* brings together ideas of the potency of the gaze—the all-seeing-eye—and the power of its ubiquity when the subject is unable to recognize what is observing it, yet knows it is being observed. The undead Count, in particular, through his inherently transformative nature, exemplifies how the same gaze is able to change and disperse itself across many platforms/devices and is able to enact forms of punishment when the subject refuses to be socialized—which is seen in the figures of Harker and Renfield. As such, the vampire embodies something of a convergence of modes of surveillance, where convergence is "the merging of roles between the actors and processes in the surveillance game, even when the aims of surveillance are different (monitoring populations or individuals in the case of the state, of the public space of the neighbourhood, or the operations of the state in the case of the individual)" (Nayar 2015, p. 7). Consequently, even the technologies of vampiric (undead) surveillance can turn—punish—the subject by turning them into a vampire. This is exampled in the plot of the next narrative to be considered—*Afflicted* by Derek Lee and Clif Prowse.

DEATH IS NOT THE END

Afflicted is a film predicated on self-surveillance. It tells of two Canadian friends, Clif and Derek, who decide to go on the trip of a lifetime around Europe and they discover that one of them, Derek, has a medical condition that might mean he can die at any moment. Clif gets a hand-held video camera to take with them so they can film the entire trip and upload all the footage to the Internet to form an online video blog, so that their friends and families can observe what they are doing. Their trip oddly parallels the one made by Jonathan Harker in *Dracula*, where the young solicitor travelled from the civilisation and privilege of the British Empire to the "land beyond the forest" (Stoker 1996, p. 266). There, these qualities no longer carry any currency, and his arrogance saw him ignore local customs and the "law" of vampiric surveillance. Clif and Derek, though coming from North America, follow this same pattern. As observed by Aaron Michael Kerner in relation to the film *Hostel* (2005) by Eli Roth, which shares a similar premise to *Afflicted* of showing two North Americans, Josh and Paxton travelling in Europe:

> Josh and Paxton [who equate to Clif and Derek in *Afflicted*]…are clearly privileged and arrogant and believe the world is theirs for the taking—and more specifically that European women are theirs for the taking. Despite all of this, and no matter how book smart they are, they are not particularly attuned to their environment. (2015, p. 107)

Clif and Derek do much the same during their trip, and although they stay in relatively well-known cities to begin with, it is when they venture beyond the tourist areas that they become caught up in a situation they can no longer control, get "bitten" by the gaze of the vampire and so become subject to the same rules of surveillance exampled in *Dracula*. Here, then, the vampiric eyes are the hand-held camera and the online blog (acting as a technology of social surveillance), and these are the devices that ensure compliance through the knowledge of being subjects of observation. The use of the hand-held camera is particularly interesting in this narrative as it places us directly behind the lens of the vampiric eye, stalking its victim and ready to condemn any actions beyond those allowed in the new environment. This notion of the audience being somehow implicit in the regulation, and eventual punishing, of the transgressive body is heightened by the use of the hand-held camera, as such films, as noted by Caetlin Benson-Allot, "presents itself as a pure *object trouvé*. In short it draws on a long history of found footage film making to trick its viewer into accepting it as 'ready-made' or real" (Benson-Allot 2013, p. 172). This is further qualified by Alexandra Heller-Nicholas who comments that such films question "our confidence that we know where the lines between fact and fiction lie are directly challenged" (Heller-Nichols 2014, p. 4), which only adds to the feeling of the audiences' complicity in the act of observation, and the inevitable vampiric consequences of transgression. Consequently, the camera lens is an extension of the gaze of the audience, just as Dracula's castle is a part of his.

Returning to the film itself, the audience, via the camera lens, has been involved in all the events of importance happening to the two main protagonists until they go to a gig in Paris held by some friends of theirs in a band called Unalaska. Here Derek begins to speak to Audrey, who we later discover is a vampire, whilst Clif encourages him to try and have sex with her, proposing this because they are in a foreign place where they think there will not be any repercussions, and indeed, all of the time the camera is on him, things seem fine. Clif and a couple of friends decide to burst in on Derek and the girl mid coitus, but when they do so, the girl has already gone. Derek is passed out on the bed with wounds on his arm and head, but with no memory of what happened. The fact that Derek was not being filmed during the incident, and that he has no memory of it, creates a dead spot in the narrative of the film, or a gap in the surveillance, suggesting that the dramatic changes to Derek occur when he is not being filmed/observed. It is as though the vampirism of the lens can only occur when it is not observing the subject of its gaze, or rather the subject is punished whenever it tries to escape the vampiric panopticon. Subsequently, the consequences of this punishment are always displayed in front of the lens afterwards. This finds correlation in Browning's version of *Dracula* where, although we see the Count and his wives observing Renfield, we never see the moments when he is bitten (punished) by the vampiric gaze. One moment, in particular, shows this as the three brides of Dracula advance on the bemused Renfield, the Count suddenly appears—after transforming from a bat back into his human form—and warns the women off with the admonition that "he is mine." However, as the vampire leans over the now supine Renfield to enact his punishment, the camera fades and the scene ends. This tableau is almost the exact mirror image to Foucault's formulation now seeing the sovereign gaze exacting its punishment in darkness, leaving its results for the light of day.

The friends leave Paris and move on to Italy, but Derek does not feel well and spends several days sleeping—there is some equivalence here to Harker/Renfield as mentioned earlier. When Clif finally gets him to go out for a meal, he is violently sick. They go to a vineyard after this, but the sunlight causes Derek's skin to blister, and when he attempts to escape, he punches a hole in a wall. Surprised by this sudden extreme strength the boys experiment with Derek's newfound superpower, neglecting to question the nature of the changes he is undergoing. As mentioned previously, Derek's increasing vampirism is intimately linked to surveillance, both through the hand-held lens that observes him, but also when he is not being observed by it, suggesting that his body is, in fact, undergoing changes due to a process of self-surveillance. This suggests that his very substance is so attuned to being watched that it enacts its own form of punishment when it is not being surveilled. There is also an element of this that sees Derek's changing or disintegrating identity being directly linked to the surveillance technologies that are watching him. Sebastien Lefait, in his book *Surveillance on Screen*, notes the impact of "new technologies on identity formation and deformation" (Lefait 2013, p. 161), and the vampiric gaze inherently initiates and propels such changes on all who come under its scrutiny. The same effect is seen in Dracula where the new technologies of the phonograph are seen to control and change the identity of those it describes. As such, the process of being observed by the vampiric, or undead gaze, of these undead technologies de-forms, or deconstructs, the subject's identity and rebuilds it in its own image—the subject becomes themselves undead or vampiric, one who surveils themselves so that they remain neither alive (authentic) or dead.

This process continues as the film develops, seeing Derek being able to leap great heights and developing an increasing thirst for blood. The most extreme example, and the one that shows the gaze has transformed him beyond any chance of returning to his former unsurveilled—original or authentic self—state, is when he tries to resist his increasing transformation and shoots himself in the head with a gun. Although he blows the back of his head off, his identity formed under the scrutiny of the lens reforms itself, and he comes back to life (undead life), and in some ways, intimating the immortality of one's persona online and in data bases, even beyond the dissolution of one's "real" body. As noted by Robert Shimonski, "Once you put something on a server such as a blog post, a data file, or other source of data, it can stay there a long time, possibly forever" (Shimonski 2015, p. 55), potentially making one's "observable" identity immortal, or one might say, undead, just like Derek's vampiric self. It can be argued that Dracula himself is made immortal through the recording device of Stoker's book maintaining his observable vampire self. During this process of vampiric change, the two friends decide to try and reverse the gaze and track down Audrey, who they hold responsible for Derek's new identity as a vampire. Derek's behaviour gets more erratic, and he leaves his friend, taking the camera (the vampiric lens) with him. Returning to France, Derek finds a phone number for Audrey and arranges to meet her. At the rendezvous, Derek hides to observe her arrival, but a man appears in her place, whom he then follows to a seemingly derelict building. Once there, a team of military dressed individuals raid the building who shoot Derek and leave him for dead. However, the bullets have not killed him and, as seen earlier, he heals extremely quickly and starts to attack the men. Derek's vampiric conversion seems complete as he roars and howls, throwing the men around the building, systematically killing them all, some by ripping their throats out. The importance of the vampiric gaze to Derek's new identity is stressed by the camera that he is still carrying with him, and the light it shines, as it is the only light in the otherwise pitch black space. It gives the effect that his (our) eyes light up anything they look at, and because of the extreme violence he enacts on the raiders, it seems the source of their punishment as well—subjects only exist within the gaze of surveillance and it decides who lives and who dies. Derek escapes the building and receives another text from Audrey. This time they meet, and Derek insists she tells him the cure, but she says: "this is not a disease, there is no cure" (*Afflicted* 2013). They fight and Derek pleads with her to kill him, but when she plunges a wooden stake into his heart, he revives. Audrey tells him that if they could be killed, she would have done so to herself—again pointing to the potential indestructibility of one's identity once it has been created by undead technologies. She also warns him that he must feed every four to five days, or he will become something worse—an uncontrollable monster that kills relentlessly. This last part symbolizes the law of the vampiric gaze that must feed itself by holding the subject in an unchanging, undead state, and to transgress that exacts severe (sovereign) punishment, which the vampire body will enact upon itself. All of the footage that the audience has seen has been uploaded to the original blog set up by Clif to chart their trip—marking a parallel to Mina in *Dracula,* who also collates and stores the various bits of undead information. In one final message to his parents, Derek tells them he can never come home, but that he had found a way to feed his new identity, and that is by killing social outcasts to survive. To that end, the film finishes, showing Derek turning to his latest victim who is tied up behind him, a character categorized as a paedophile, due to material recorded on his phone—another form of surveillance. Once again, the only light provided is from the camera and so the vampiric gaze of our observing eye alights, and feeds on its latest victim.

ALL CONSUMING GAZE

Daybreakers contains elements of the earlier two films, showing the nature of the vampiric gaze both on society and the individual, but more so than either of them, it displays the all-consuming nature of surveillance; how the appearance of choice can be no choice at all.[2] The film is set in the near future, 2019, after there has been a pandemic of sorts that has turned everyone into vampires. This is not unlike Richard Matheson's seminal novel *I Am Legend* from 1954, which depicted the first such worldwide vampire plague, except in 2019 the infected are not the shambling vampire/zombies, or "Zampires" as Victoria Nelson calls them (Nelson 2012, p. 154), but not unlike the human population they have replaced. As noted by Gelder, "it seems as if vampires have exactly copied humans, or the future of humans; but their increased life expectancy, the thing that humans normally aspire towards … is represented here as an increasing drain on what sources remain" (Gelder 2012, p. 130). The resources here, of course, are humans and what few there are left are captured and placed into blood farms—not unlike battery chickens—to be slowly bled to feed the few that can afford it. For those unable to get enough blood, or decide to feed on themselves, the result is something not unlike that described to Derek in *Afflicted*, but here, as Paul Meehan comments, "unfed vampires degenerate into 'subsiders,' bald, bat-winged mutants who lose all capacity for rational thought" (Meehan 2014, p. 87). The restriction of society manifested in the lack of blood to eat is mirrored in the vampire's surroundings where, as Meehan further notes, "this vampire society inhabits a dark city, where everyone hides from the sun and wears black clothing" (Meehan 2014, p. 87). The vampire's extreme sensitivity to sunlight means they live in a totally enclosed space, office buildings—all have UV resistant glass—and transport systems are linked by tunnels, and even vehicles are completely blacked out, relying on video feeds of the outside world to drive safely. Yet for a society so enclosed, the feeling is that it is constantly being surveilled, either by CCTV, or by one's fellow citizens, not unlike the inhabitants of the land around Dracula's castle. Indeed, in many ways, the film shows the future that Stoker's vampire envisioned for the British Empire once he made it his own. This impression of being under continual observation is enforced through the many scenes showing video feeds of crowds waiting for trains, and prisoners in their cells. The continuous observation is also evident in the constant anxiety of the film's main protagonist, Edward Dalton, over being discovered transgressing the rules of the vampire society.[3] The panopticon here works in two ways, firstly through the self-surveillance of the society itself—all are encouraged to report anyone not conforming to the rules of blood rationing—and secondly, through the very architecture of the observational space they are forced to exist in. In this way, the vampire population and the fabric of the space they live in becomes one self-surveilling body, regulating itself and all who pass through it. This all-consuming vampiric gaze also repositions the meaning attributed to the "subsiders" who, unsurprisingly, live in tunnels and parts of the city that are not observed. However, when they re-enter the field of the gaze, they are presented as shocking, grotesque examples of the punishment of transgressing the rules of self-surveillance. Not unlike Derek in *Afflicted*, it is when the surveilled body leaves the field of observation—when the subsiders retreat to their underground lairs—that it undergoes its transformation into something monstrous. *Dracula* constructs a similar system, as it is at night, in darkness, that the punished body undergoes its most dramatic changes and is then displayed showing its new vampiric status. The final punishment of the vampiric gaze is, then, to banish transgressive subjects from ever being seen again, which is performed by chaining the subsiders to a large vehicle and then dragging them out into the sun where they burst into flames and disintegrate, achieving a complete and total deformation/destruction of the identity created in the vampiric lens. This is the aspect of *Daybreakers* that marks it out as the culmination of the kinds of identity formation under surveillance seen in the previous examples.

Dracula showed an environment where the behaviour of the subject was controlled by the all-seeing vampiric eye, in all its forms, but those being observed were able to leave. The ultimate punishment for transgressing the rules of being surveilled by the vampire was having your old identity deconstructed and replaced by one formed in the eye of the undead observer. In *Afflicted* a similar construction is seen, except that once the identity formed in the vampiric lens has usurped that which the subject had before, it can never be destroyed or replaced—though as mentioned earlier, the narrative of *Dracula* itself ensures a level of immortality to the vampire. Interestingly here, although in *Dracula* and *Afflicted*, there are multiple devices recording/observing the subject, potentially allowing for what Maria Los sees as "multiple surveillance orientated 'looking-glass selves'" (Los 2006, p. 82), the identity/ies created are extremely uniform and constant. This uniformity finds its apotheosis in *Daybreakers* where the entire society is comprised of vampires who all have the same identity construction and are self-surveilled by their own bodies and their environment to remain that way. Alternatively, their bodies turn them into a subsider, which is then ejected by their environment. The range of observation devices become the source of an immortal, or undead body that creates a single identity position for the subject for the entirety of its existence.

Daybreakers actually offers some form of release from this totalitarian identity machine, when the hero Edward Dalton manages to change his vampiric identity to something else. His new identity is not vampire and not quite the human he was before, but something else—whose blood has the effect of altering any vampires that drink it. Dalton gives his blood to Charles Bromley, the figure that runs the blood farms, and once he has changed from being a vampire, offers him to his own security forces, who instantly attack and kill him. But they then change from being vampires and are themselves attacked by other members of the security force, and, as noted by Gelder, "an eternal cycle of consumption is played out, potentially without end" (Gelder 2012, p. 132). However, just as this seems to be getting out of control, a scientist who Dalton used to work with, runs in and shoots the non-vampires, exclaiming that there cannot be a cure, before being killed himself. Yet Dalton and his accomplice Elvis, who are both non-humans, are still alive, suggesting that there is a potential to break out of the unchanging and unchangeable undead subject position created by the vampiric lens, should they choose to do so.

CONCLUSION

This study has considered the use of surveillance within films that feature vampires that are created, exist, and even die within the "vampiric" undead gaze of observation. Whilst the examples used are fictitious, they talk of the possible and actual effects of surveillance upon individual and social identity creation, as well as notions of inclusion, exclusion, self-governance and even self-punishment. Films and novels like *Dracula* might seem far removed from the concerns of the early twenty-first century, but they do capture something of the gothic nature of surveillance—a ubiquitous unseen ghost. *Afflicted* purposely complicates the boundaries between the real and the imaginary to make its critique on observation more immediate to show equivalences between the undead observational devices from *Dracula*—phonograph, telegram, photograph—with those of the Internet age: hand-held cameras, blogs, social media. *Daybreakers* takes this to its logical, inevitable, conclusion showing a world in the near future where surveillance is so total and ubiquitous that it is part of the very nature of the environment. This reveals the truth of William Bogard's observation that "power [as symbolized by the panopticon] disguises itself by disappearing into the architecture" (Bogard 2012, p. 31). In this vampiric society, the undead eye of Dracula's castle no longer watches the world; it *is* the world. *Daybreakers*, in particular, constructs this as a simultaneously global, but hermetic world, where the fear of losing what one has is directed by a central, surveilling power. This anxiety is further exacerbated, at least in terms of the film, by labelling humans—those who still possess an unsurveilled identity—as enemies of the state that need to be tracked and hunted down, an endeavour that all citizens are encouraged to be part of. These fictional conditions oddly parallel those of the early 2000s where, as noted by Armand Mettelart:

> Failing or refusing to have one's personal data recorded is the crime today. The tracking grid now provides meaning on a planetary scale, even though it takes on the contours of each particular reality...Powerless to combine aspirations for freedom with aspirations for security the heads of the world system network resort to strong-armed management of inequalities instead of declaring war on the mechanism that reproduce them...These realities are only intelligible when examined in relation to the combined effects of capitalists unbridled project to control every facet of life and the revolts this project stirs up against itself (Mettelart 2010, pp. 2–3).

The combination of security and consumerism is a prevalent one, not unlike that practiced by Dracula himself, whose vampiric gaze created a growing society of consumers, a project which finds its fulfilment in *Daybreakers*. Alongside this are the means and devices of surveillance which, theoretically, provide the possibility for the creation of multiple selves and a means of resistance to globalised/controlled identity positions. However, as seen in *Afflicted*, and even more so in *Daybreakers*, these "other" positions can themselves be unchanging and unchangeable (undead) identities that, once beyond the control of the subject, can live forever.

NOTES

1. The importance of the vampiric gaze through the highlighting of the eyes is seen in its use in later films such as *Dracula's Daughter* by Lambert Hillyer from 1936 and, more recently, Morticia Addams, from the Addams Family, both the television series (1964–1966) and the later films, *The Addams Family* (1991) and *Addams Family Values* (1993).
2. Whilst *Dracula*—both book and film—and *Afflicted* share something of a post-colonial thematic where the naïve representative of an Imperialist empire travels to the "Old World" and gets/brings back more than they bargained for—the previously consumed becoming the consumer, *Daybreakers* is more self-contained showing a nation that is consuming itself. Resultantly it speaks more of the exploitation of Australia's limited resources than the relations between aboriginal, colonial, and later immigrant communities.
3. Edward no longer wants to drink human blood and has started to show telltale symptoms of his body punishing itself for doing this, i.e. he is slowly turning into a subsider.

REFERENCES

Abbott, S. (2007). *Celluloid vampires: Life after death in the modern world*. Austin: University of Texas Press.

Afflicted. (2013). Directed by Derek Lee and Clif Prowse. CBS Films.

Astle, R. (1979). Dracula as totemic monster: Lacan, freud, oedipus and history. *SubStance, 8*(25), 98–105.

Benson-Allot, C. (2013). *Killer tapes and shattered screens: Video spectatorship from VHS to file sharing*. Berkeley: University of California Press.

Bogard, W. (2012). Simulation and post-panopticism. In K. Ball, K. D. Haggard, & D. Lyon (Eds.), *Routledge handbook of surveillance studies* (pp. 30–37). London: Routledge.

Daybreakers. (2009). Directed by the Spierig brothers. Lionsgate.

Dracula. (1931). Directed by Todd Browning. Universal Pictures.

Foucault, M. (1977). *Discipline and punish: The birth of the prison* (A. Sheridan, trans.). New York: Vintage.

Gelder, K. (1994). *Reading the vampire*. London: Routledge.

Gelder, K. (2012). *New vampire cinema*. London: Palgrave Macmillan.

Heller-Nicholas, A. (2014). *Found footage horror: Fear and the appearance of reality*. Jefferson: McFarland.

Hutson, L. (2005) Rethinking the 'spectacle of the Scaffold': Juridical epistemologies and English revenge tragedy. *Representations*, No. 89, 30–58.

Kerner, A. (2015). *Torture porn in the wake of 9/11: Horror, exploitation and the cinema of sensation.* New Brunswick: Rutgers University Press.

Leblanc, B. (1997). The death of dracula: A darwinian approach to the vampire's evolution. In C. M. Davison & P. Simpson-Housley (Eds.), *Bram stoker's dracula: Sucking through the century, 1897–1997* (pp. 351–376). Toronto: Dundurn Press.

Lefait, S. (2013). *Surveillance on screen: Monitoring contemporary film and television programs.* Lanham: The Scarecrow Press.

Los, M. (2006). Looking into the future: Surveillance, globalisation and the totalitarian potential. In D. Lyon (Ed.), *Theorizing surveillance: The panopticon and beyond* (pp. 69–94). London: Routledge.

Lyon, D. (Ed.). (2006). *Theorizing surveillance: The panopticon and beyond.* London: Routledge.

Meehan, P. (2014). *The vampire in science fiction and literature.* Jefferson: McFarland.

Mettelart, A. (2010). *The Globalisation of surveillance: The origin of the securitarian order* (S. Taponier & J. A. Cohen, Trans.). Cambridge: Polity Press.

Nayar, P. (2015). *Citizenship and identity in the age of surveillance.* Cambridge: Cambridge University Press.

Nelson, V. (2012). *Gothicka: Vampire heroes, human gods, and the new supernatural.* Cambridge: Harvard University Press.

Pick, D. (2000). *Svengali's web: The alien enchanter in modern culture.* New Haven and London: Yale University Press.

Rhodes, G. (2006). *White zombie: An anatomy of a horror film.* Jefferson: McFarland.

Shimonski, R. (2015). *Cyber reconnaissance, surveillance and defense.* Waltham: Elsevier.

Slobogin, C. (2007). *Privacy at risk: The new government surveillance and the fourth amendment.* Chicago: University of Chicago Press.

Stoker, B. (1897). *Dracula.* London: Signet Classics, 1996.

Author Biography

Simon Bacon is an independent scholar based in Poznan, Poland. He has lead interdisciplinary research projects on Vampires, Myths and the Past and Future (November 2011, London). He was editor in chief of the Monsters and the Monstrous Journal from 2011–2014 and has co-edited books on *Undead Memory: Vampires and Human Memory in Popular Culture* (2014) and *Little Horrors: Interdisciplinary Perspectives on Anomalous Children and the Construction of Monstrosity* (forthcoming).

PART II

Literature

CHAPTER 7

Watching Through Windows: Bret Easton Ellis and Urban Surveillance

Alison Lutton

"Discovering Japan", one of the most widely known short stories from Bret Easton Ellis's 1994 collection *The Informers*, both opens and closes with its protagonist, obnoxious American rock star Bryan Metro, staring (or attempting to stare) out of an aeroplane window, surveilling the vast urban landscape of Tokyo. At the story's outset, Bryan, representative of a stereotypically detached early Ellis protagonist, cannot discern his location due to wearing sunglasses at night. At its close—and having made numerous problematic intrusions into overlooked urban space—Bryan is once again gazing through a plane window, and this time notes that "I relax for a moment when the lights of Tokyo [...] vanish from view but this feeling only lasts a moment because [...] other lights in other cities, in other countries, on other planets, are coming into view soon" (p. 129).

This wholesale shift in perspective on Bryan's part—from literal, blank ignorance of his orientation vis-à-vis the urban environment he is approaching, to a detached, but all-seeing appreciation of this terrain from both a personal and transnational viewpoint, is representative of the simultaneously problematic, surprising, and self-reflexive encounters with urban space experienced by each protagonist of the thirteen short stories that constitute *The Informers*, from a whimsical young girl casually and inevitably slipping into an increasingly superficial LA lifestyle, related through letters back home to an unrequited love on the East Coast in

A. Lutton (✉)
University of Oxford, Oxford, UK
e-mail: alison.lutton@some.ox.ac.uk

© The Author(s) 2017
S. Flynn and A. Mackay (eds.), *Spaces of Surveillance*,
DOI 10.1007/978-3-319-49085-4_7

"Letters from L.A.", to vampires terrorising the city in "The Secrets of Summer". Bryan's predicament is, moreover, illustrative of another key feature of each of *The Informers'* narratives: the surveillant tendency, generally manifested via a preoccupation with both seeing—or, more accurately, observing—and being seen, but often communicating something deeper about the pervasiveness—and potential pleasure—of surveillance in the contemporary urban environment. While, during his time in Tokyo, Bryan is unable to fully see his audience as he performs onstage (p. 117), or himself as he stares into a mirror while—again—wearing sunglasses (p. 127), his understanding of himself as a subject of surveillance within the urban environment is never in doubt: "[w]e pass a crowded street diagonal to the hotel and everyone is trying to peer into the tinted windows as the limo rolls toward the Hilton", he at one point recounts, an experience which prompts a reaction of both awe and revulsion, and which foregrounds some of the many windows through which characters see— and are, perhaps, seen—in *The Informers* (p. 118).

The effect of this tendency relates to what Georgina Colby, to date one of only a few critics to engage with *The Informers* in depth, in a reading of the prevalence of the bystander in the collection, has identified as the way in which its deployment of the "ubiquitous blank gaze [...] signifies at once the absence of the subjects and a form of desire" (2011, p. 57). This reading of *The Informers* will focus on drawing out the desire of this apparently blank bystander, certainly a prominent figure, but one who also communicates much of significance about both their own identity and the often claustrophobic and overlooked urban environment which they inhabit through their practice of, and subjection to, surveillance—they may be gazing blankly, but they are often also seeing, (and sometimes, further, being seen). For Colby, the desire in question is "not for the other but for the event" (ibid.). Here, Colby is specifically invoking the transformative potential of the Badiouian event, as reflected in Ellis's documentation— and simultaneous implicit rejection—of the worst excesses of the 1980s (ibid., p. 29); I would suggest, in response to this reading, but on a tangential note, that with regard to the collection's surveillant tendency, a further key event arises from the amalgamation of overseen subjectivity and the surveilled and surveillant urban environment in which this is taking place, which reveals much about both.

This event, as we will see, is uniquely rendered in *The Informers*—and further critical opinion on the collection suggests it to have a potentially pivotal position within Ellis's canon. Colby, following Julian Murphet, who, writing in 2002, found *The Informers* to be "the most accomplished text of Ellis's career" (2002, pp. 17–18), views the collection as a key component of Ellis's three earliest works, which, she feels, should be valued as important elements of his canon in their own right (2011, p. 25). Sonia Baelo-Allué, meanwhile, has also found that the text, published between the vastly successful *American Psycho* (1991) and the appearance of Ellis's subsequent novel, *Glamorama*, in 1998, resonates strongly with Ellis's earlier work— and, particularly, with his debut, *Less Than Zero* (1985)—representing an apparent "return to the style and fictional world" of that novel (2011, p. 131) —a clearly evidenced point. However, Baelo-Allué subsequently makes the compelling argument that, while Ellis's critical reputation had been cemented at an early point in his career, he has nevertheless "followed a logical evolution in the subjects he has chosen and in the way he has dealt with this material" (p. 133). With this reading falling within a chapter which more broadly considers *Glamorama*'s reception, *The Informers* performs an important role in directing negotiation of that text, and further emerges as a potentially vital constituent part of Ellis's canon. As Baelo-Allué highlights, *Glamorama* (a text which, notably, extensively depicts transnational surveillance [and on its transnationalism, see Lutton 2013]) features within what might be most succinctly identified as a continuum running through Ellis's fiction, characterised by narrative and thematic currents, which both draw on, and progress from, his textual past (pp. 133–135). Baelo-Allué strongly implies that *The Informers* forms an important constituent part of this continuum (progressing, for example, to stress the relatively—although differently—"plotless" nature of the two texts [p. 140]), but does not explicitly make a case for this. I would now like to make this case, arguing, in addition, that the collection's thematisation of surveillance forms a crucial aspect of its construction within this continuum.

To facilitate deeper exploration of the pivotal position of *The Informers* within Ellis's canon, it is necessary to situate more clearly its realisation of its surveillant urban environment. The way in which the collection creates, and shows its protagonists to inhabit, mid-1980s Los Angeles, is crucial here. A fitting analogue of the kind of nuanced, interrelational depiction of the city in the collection can be found in Scott McQuire's theorisation of the turn-of-the-twenty-first-century "media city", a space dominated by radically new "spaces and rhythms", created through "a complex process of co-constitution between architectural structures and urban territories, social practices and media feedback" (2008, p. vii). While McQuire, writing in 2008, is specifically imagining the post-digital city, the relationship he invokes—particularly when discussing "relational space" in the urban context (p. ix)—can very much be seen to be exemplified in the varied relationships between urban territory and both social situatedness and mediated feedback of many types lying within the pages of *The Informers*.

A deeper look at McQuire's formulation of the media city—or, as he also terms it, a "*media-architecture complex*" (p. vii)—reveals this work to be particularly applicable to a reading of Ellis. McQuire is quite explicit in his association of the contemporary city with both transnational media flows and, importantly, with the associated, encroaching inevitability of surveillance, finding that:

> Contemporary cultural identity is […] less defined by an "imagined community" based on the geographical borders of a single national territory, but increasingly assumes the mosaic pattern adumbrated by the overlapping footprints of satellites and the flows of digital networks. At the same time, media devices have become ubiquitous elements shaping the public spaces of contemporary cities, embedded in urban infrastructure in a wide variety of locations and forms from informational kiosks, large public screens, digital surveillance cameras and computerized traffic systems. (p. 6)

In finding the media city to be arranged in a "mosaic pattern" incorporating varied constituent parts, both local and transnational, McQuire calls to mind the notion of the assemblage, a key concept in recent theory written under the umbrella of surveillance studies, and popularised by sociologists Haggerty and Ericson (2000). The assemblage denotes the ways in which ostensibly differing modes of surveillance work alongside each other, particularly in the context of the contemporary urban landscape (Lyon, pp. 98–99), and which can be understood to be "comprised of heterogeneous component parts which are aligned through processes of disassembling and reassembling" (Lyon et al. 2012, pp. 15–16). The naturally varied and complex nature of assemblages means that, according to Lyon, Haggerty and Ball, "it is extremely difficult to pinpoint the locus of responsibility for surveillance processes" (p. 9) when they are deployed —clearly, then, the complexity and interrelationality of the urban space of surveillance cannot be underestimated.

The two key features of assemblages identified above—the inevitable process of disassembling and reassembling, and the difficulty they present in pinpointing individual agency—can be seen to be highly—and, therefore, presciently (given that the concept of the digital urban assemblage was not fully active, much less understood, until the 2000s)—evident at a tonal level in the literary fashioning of the mosaical media city of Los Angeles by Ellis in *The Informers*, with the agency of characters deconstructed, and subsequently reconstructed, against the surveillant city space which surrounds them, and personal agency, as might be expected from Ellis, problematic, but also reflexively engaged with, and through, the practice of surveillance. With this in mind, the rest of this chapter will explore *The Informers* from two key perspectives: firstly, as fully stressing Los Angeles, for the first time in his canon, as a media city, a site of surveillance and a space of assemblages, of the overlooked, but also of the overlooking; and secondly, as anticipating the representations of city spaces as full, turn-of-the-century media cities, which would come to predominate Ellis's writing post-1998—particularly, in his most recent, and, significantly, LA-based, novel, *Imperial Bedrooms* (2010).

Imperial Bedrooms is an important successor to *The Informers*, reprising, in the twenty-first century, the collection's representation of the surveillant space of Los Angeles, and providing an excellent example of how such mediated and overlooked city space impacts on conceptualisations and representations of individual agency in Ellis's later fiction—particularly, those concerned with disruptions to everyday life caused by the surveillant assemblages, immediate or remote; as such, it is worth dwelling on in more detail at this point. The novel notably engages with core concepts of surveillance theory, as defined above, in a most important way: it explicitly illustrates what McQuire has defined as the increasingly porous boundary between the homestead and the city space in which it is set (2008, p. 6), emphasising, as is most evident in comparison with a similar, and yet jarringly different, way of life in Los Angeles depicted in *Less Than Zero*, to which *Imperial Bedrooms* serves as an effective sequel, the way in which increasingly pervasive surveillance—or "surveillance creep", as it has been termed (Lyon 2007, p. 52)—is an important factor in the unsettling of earlier rhythms of everyday life, a primary effect of the genesis of the media city, as McQuire would have it (2008, p. 6).

This designation is especially relevant to the experience of Clay, protagonist of *Imperial Bedrooms*—and, formerly, of *Less Than Zero*. In the former narrative, the Los Angeles of 1985 has shifted into a media city of the twenty-first century, arguably undergoing a far greater evolution than any of the novel's recurring characters. In this space, Clay is forced to renegotiate his own experience of the homestead, which was previously, in *Less Than Zero*, a site of disenchantment, but largely not an overlooked space, which allowed for some—albeit blank, disengaged, and ultimately dissatisfying—respite from the alienation and underlying sense of dread he experienced on the streets, and in the bars, restaurants, and parties of LA, where anyone might "Disappear Here". Twenty-five years later, Clay inhabits his new urban homestead very differently, due to paranoia that its privacy has been invaded; he is constantly aware of the porous boundary between home and external urban space, and of his own potential subjection to surveillance even behind closed doors. A typically paranoid negotiation of this private space reads thus: "Moving around the condo I ask, 'Is anyone here?' I lean over the duvet in the bedroom. I run my hand across it. It smells different. I check the door for the third time. It's locked..." (2010, p. 25). Clay's sense of being overlooked is, further, exacerbated by the prevalence of new media technologies (particularly, the mobile phone through which he receives increasingly sinister text messages despite being physically safely ensconced within the walls of his condominium). At one point, for example, Clay, home alone, receives some typically unsettling digital correspondence, which powerfully reconfigures his occupation of ostensibly private space: "I glance out of the window again and am surprised when I find myself backing into a wall. The condo suddenly seems so empty but it isn't—there are voices in it" (p. 51). The porous boundaries of the twenty-first century media city here, evidently, figure Clay as an increasingly overlooked and threatened subject.

It is clearly significant that it is only through the increasing prevalence of technology that the all-pervasive mood of surveillance established in *The Informers* is able to be fully realised by Ellis; this is, again, suggestive of the progressive tone of that text, which began to exercise the shifting mood of the city long before its architecture and infrastructure had provided the means to fully illustrate this, as it had by the publication of *Imperial Bedrooms* (which is, incidentally, the only work of fiction by Ellis to be set entirely in Los Angeles since *The Informers*) some sixteen years later. This fact is equally significant at an off-textual level, with the idea of the home also being problematised for Ellis's longstanding readers, who had come to associate Clay and the narrative of *Less Than Zero* with residence in a particular kind of Los Angeles, culturally embedded in the coordinates of the high 1980s sensibility, which it is widely understood to reflect. For the reader (himself necessarily a surveillant presence) to be confronted with a Los Angeles, written by Ellis, but transformed into a media city of a particularly turn-of-the-century variety, as imagined by McQuire, doubles the unease associated with the highly mediated and overlooked potential of the contemporary urban landscape. The unsettling of the reader here mirrors the jarring of the reader initiated in 1994 with the publication of *The Informers*, which, in its narrative tone, at times imagines a markedly more sinister Los Angeles than that represented in *Less Than Zero*; in *Imperial Bedrooms*, the potential of *The Informers* is fully realised through the textual emplotment of the kind of city-based assemblage of technologies and symbols of surveillance described by McQuire. As J. Robert Lennon notes, contrasting the novel's cultural coordinates with what Ellis's reader has come to appreciate as the highly familiar 1980s ephemera of *Less Than Zero*, "[h]ere, we [instead] get iPhones, Apple Stores, internet videos [...] all of them crated up and ready to go on loan to the Historical Archive of the Post-Bush era" (2010, p. 22). With this in mind, *The Informers* can be seen to anticipate the mood and mediation that would shape the literature of surveillance Ellis would progress to write, creating an impression of the surveillant city which technological detail would subsequently allow to be refined.

With *The Informers*' textual legacies in mind, it is now most useful to return focus to that text, and to examine more closely the genesis of the moods and spaces of surveillance described above. One particular story is most positively representative here: "Water from the Sun". Cheryl, the story's protagonist-narrator, is, fittingly for a resident of a media city, a newscaster and a (sometimes willingly) surveilled subject, who represents what Lyon defines as a tendency in which surveillance "is about vision, but not one-sidedly so; [it] is also about visibility' (2007, pp. 15–16)—and whose negotiations of her own homestead and the city space in which it is set are increasingly fraught and disorientating as the narrative progresses. The nature of her employment means that Cheryl is a character who exemplifies most literally the mood of surveillance which pervades *The Informers*—while she is initially introduced to the reader as a viewer of television, we are subsequently reminded of her status as a subject with the consistent potential to be overlooked—"[e]veryone in the restaurant, it feels, recognizes me", she notes at one point (p. 86). Further, Cheryl is herself engaged in the localised practice of surveillance as she monitors the movements of her young lover, Danny, whom she suspects of cheating. From the opening of the story, Cheryl invokes a prototypical version of the collapse of the public/private binary associated with the uncanny homestead located in the media city (McQuire 2008, pp. 6–9), expressing frustration that Danny has not taped the newscast on which she has appeared, rendering her unable to review her public, mediated performance from the comfort of her own bedroom. Cheryl's desire for active spectatorship here acts as a point of contrast with the numerous depictions of passive television viewing in Ellis's early fiction, which led to its classification as "MTV fiction" by a number of critics (on this tendency, see particularly Buscall 1999). While Danny is amongst a number of characters in *The Informers* shown to watch MTV passively (on pp. 84, 90), Cheryl is categorically not; and the introduction of this tension between ways of watching infuses this story with an important additional dynamic, again setting it apart from Ellis's earlier work.

The narrative which subsequently unfurls both situates Cheryl within the contested and overlooked spaces of a prototypical media city, and very much depicts her as a determined presence therein, both surveilled and surveillant, with the urban architecture she negotiates informing and heightening her experience of surveillance. This heightened experience is particularly apparent in Cheryl's negotiation of two key sites within this prototypical media city: the porous window onto the thoroughfare, and the airport.

The prevalence of surveillance in the story is initially made explicit through the deployment of the trope of the window, associated throughout *The Informers* with the act of concealed spectatorship, or personal surveillance. During a lunch with her agent, Sheldon, Cheryl initially finds it "distressing" that her companion repeatedly stares out of the window at a palm tree; casual and yet active spectatorship here establishes a threatening tone (pp. 85–86). However, it is when Cheryl herself looks through the same window, ultimately focusing on the very same palm tree, that a particularly notable—and sinister—mood of surveillance is established:

> I'm looking out the window again, at two teenage girls with short blond hair, wearing miniskirts, who are walking with a tall blond boy and the boy reminds me of Danny. It isn't that the boy looks exactly like Danny—he does —it's more [...] the way he checks himself out in the window of this restaurant, the same pair of Wayfarers. And for a moment he takes off his sunglasses and stares right at me even though he doesn't see me and [...] puts his sunglasses back on and makes sure they are not crooked and turns away and walks down Melrose and the two girls leave the palm tree and follow the boy.

"Know him?" Sheldon asks (p. 87).

Opening with an indication of the practice of surveillance as an unspoken, repetitive compulsion, as Cheryl looks out of the window "again", this passage foregrounds several key features of the story, of the collection as a whole and, at times, of Ellis's entire oeuvre. Firstly, it reflects the collection's straddling of tone between Ellis's early and late fiction in its description of characters' physical appearance. The watched subjects here are blondes, and one is wearing Ray-Ban Wayfarers—both key stylistic tropes of the almost interchangeable youths who peopled *Less Than Zero*. However, one of these people, at least, is (apparently) not interchangeable, invoking the figure of the doppelganger, a key presence in Ellis's novels from *American Psycho* onwards. Here, it is the uncertainty of the identity of the watched subject, which renders this act of surveillance particularly unsettling; the query "[k]now him?", which closes this section of the story, generates a jarring sense of both familiarity and distance, anticipating the uneasy relationship between self and urban environment which becomes increasingly dominant as the story continues. Secondly, the casualness with which the boy who may, or may not, be Danny situates and sees himself within what is a familiarly constructed urban environment is particularly telling. Viewing his reflection in the restaurant window, the boy momentarily rests against the palm tree which, although itself a natural object, is a highly recognisable feature of the urban landscape of Melrose Avenue, on which the restaurant is located—a feature which has been described as contributing to the creation of a city space characterised by a "veneer of bliss" (Benton 1995, p. 48)—a veneer which, as readers of Ellis are all too aware, is easily cracked. The casual way in which the boy occupies this space, using his sunglasses both as a superficial fashion accessory and a tool to shift between seeing and being seen, underlines the surveillant and surveilling potential of the everyday, which would ultimately be fully realised in the contemporary media city described by McQuire. Finally, and most notably, the dual role of the window—as a porous lens through which Cheryl sees the boy, and in which he sees himself reflected, but through which he threatens to (but ultimately, apparently, does not) see her—is strongly indicative of the kind of reciprocal dynamic identified by Lyon as an important feature of surveillance today (2007, p. 7); again, the contemporary relevance of *The Informers* is clear.

The porous, and yet sometimes curiously myopic, role of the window highlighted here is prevalent elsewhere in *The Informers*: with windows mentioned nearly seventy times throughout the collection (an average of once every three pages), the potential for characters to watch—and be watched—through glass is as consistent as it is sinister. Another notable example of this tendency comes in the story "In the Islands", which opens with the sentence "I am watching my son through a mirrored window from the fifth floor of the office building I own" (p. 42)—a clear statement of intent, replete with the sense of power which can be drawn from personal, and not just institutionalised, urban surveillance (with the cosseting nature of the homestead—or, at least, the private indoor space— troublingly reversed). While the narrator's son, Tim, repeatedly looks up at the window, the narrator is left to "stare through the glass, relieved that Tim can't see me, that we can't share a wave" (ibid.)—here, the apparently one-sided, panoptic power associated with this personal act of surveillance is actually seen to provide relief from the obligation of human interaction, creating an uneasy dynamic: the act of surveillance here reads simultaneously as eerily distanced and absolutely—and oddly positively—personal.

Given the dualistic nature of the surveillant subject established here, Lyon's underlining of the potential for positive personal engagement with practices of surveillance is again called to mind—and the unusual dynamic this evokes is reflected in another key passage in "Water from the Sun", in which Cheryl apparently aimlessly drives to the airport:

> I get into my car and [...] find myself driving to the airport, to LAX. I park and walk to the American Airlines terminal [...] and then I do some of the cocaine Simon gave me this afternoon [...] and then I roam the airport and hope someone will follow me and I walk from one end of the terminal to the other, looking over my shoulder expectantly, and I leave the American Airlines terminal and walk out to the parking lot and approach my car [...] I get the feeling that there's someone waiting, crouched in the backseat, and I move toward the car, peer in, and though it's hard to tell, I'm pretty sure there's no one in there and I get in and drive out of the airport and as I move past motels that line Century Boulevard leading to LAX I'm tempted, briefly, to check into one of them [...] to give off the illusion of being someplace else... (p. 96)

Julian Murphet, in an influential reading of Ellis's early work as situated within the racially contested, oppresively commercialised social space of Los Angeles, finds this passage to be characterised by "a frantic quest for experience and excitement which will never come" (2001, p. 81). Nathalie Vincent-Arnaud, whose discussion of *The Informers* offers an important, complex consideration of the (negative) potential of the space of Los Angeles in Ellis's prose, also views the passage as being of particular significance, identifying it as a "gigantesque hypotypose d'errance"[1] (2006, p. 109), and moreover, reads LAX as depicted in *The Informers* as a space of transit and anonymity (ibid., p. 110)—however, an even more fraught and problematic representation of the space in question emerges when it is considered primarily as a site of surveillance. The invocation of two very recognisable features of the urban geography of LA, Century Boulevard and LAX, very explicitly and deliberately problematises the narrator's self-location, creating an unsettled sense of place, which quickly gives way to a creeping (although endlessly deferred) sense of surveilled subjectivity. Of course, driving to an airport without a further destination in mind and considering checking into a motel when one has one's own private home in the vicinity, are both in themselves unexpected negotiations of particularly-purposed, and usually transitory, spaces, representing a kind of heightened experience of the postmodern flânerie evident in *Less Than Zero* (on which, see Lutton 2012), with the narrator able to repurpose the urban landscape as she traverses it. However, there is an important undercurrent here, suggesting the potential pleasure of the surveilled subject. While some of the passage's recurrent references to being watched, or followed, are arguably attributable to drug-fuelled paranoia, there are as many which actively invite the surveillant gaze—and this suffuses the unsettled description of the urban spaces in question with an additional duality.

The nature of the airport as a space of surveillance is crucial here; after all, airports are, as Lyon notes, "perhaps the most stringently surveilled sites in terms of the means of movement and identification" (2007, p. 123). For Lyon, the repressive potential of such surveilled spaces is at odds with depictions of the airport in the work of critics such as Marc Augé, which, he feels, "romanticize them" (ibid.). However, there is a bilaterality to the way in which Augé writes about the airport, which mirrors the simultaneous repression and pleasure afforded by the mood of surveillance experienced there in "Water from the Sun". In the account of movement to and through and airport, which opens his highly influential *Non-Places* (notably, an expression borrowed from Augé—in the original French, 'non-lieux'—by Vincent-Arnaud, to characterise her depiction of the space of *The Informers* (2006, p. 110), and which, in some ways, mirrors Cheryl's visit to LAX, Augé describes Pierre Dupont, his fictional exemplary passenger, as, after having checked in for his flight, "enjoying the feeling of freedom imparted by having got rid of his luggage and at the same time, more intimately, by the certainty that, now he was 'sorted out', his identity registered, his boarding pass in his pocket, he had nothing to do but wait for the sequence of events" (1995, pp. 2–3). Here, the surveillant mechanisms facilitating transit through the airport are shown to provide both freedom and control, figuring it as romanticised despite, and because of, a sense of regulation, order, and oversight stemming from its status as a space of surveillance *par excellence*. Cheryl's desire to be "someplace else", expressed in "Water from the Sun", perfectly mirrors Augé's Dupont reading, in an in-flight magazine, about "[t]he irresistible wish for a space of our own" (ibid., p. 4)—arguably, this can be found in the simultaneously liberating and repressing contemporary space of surveillance.

CONCLUSION

The surveillant space of the airport, given the simultaneously repressive and liberating "sequence of events" which, as Augé suggests, inevitably comes to pass once this space is entered, is an ideal location on which to dwell, in order to evaluate the spirit of surveillance in *The Informers* as a whole. Augé's "sequence of events" is strongly mirrored in the desire for "the event" Georgina Colby identifies in the otherwise apparently blank gaze of *The Informers'* band of detached bystanders (2011, p. 57), with which this discussion opened. At that initial point, I suggested that this event, in *The Informers*, might be extended to represent the amalgamation of overseen subjectivity and the surveilled and surveillant urban environment in which this is taking place, revealing much about both; while this is certainly the case, as we have seen, it is also clear that the event of surveillance in *The Informers* is more complex still. Neither fully constituted of blank non-spaces, nor of places of wholly institutionalised, repressive surveillance, the urban landscape of Los Angeles, as figured in *The Informers*, can be understood as a third type of space—a site of reciprocal seeing and watching, which allows its protagonists a kind of integrated relationship with their surroundings, which would anticipate the creation of spatially-imbued, constantly doubled and overlooked figures in Ellis's later work. In fact, the collection can be seen to enact its own kind of literary assemblage of surveillance, mirroring the surveillant assemblages as defined by Lyon, Haggerty and Ball, infusing many of the features of the media city as imagined by McQuire with simultaneously observant and observing, willing and unwilling participation in this, and especially stressing the collapse of the public/private binary more easily maintained in the versions of the city depicted in Ellis's earlier fiction. The event is the space of surveillance, and the space of surveillance is the event—constantly disassembled and reassembled, after Lyon, Haggerty and Ball—and in representing this, *The Informers* is both unique and eerily prescient.

NOTE

1. 'A huge, meandering hypotyposis' (translation my own).

REFERENCES

Augé, M. (1995). *Non-places: Introduction to an anthropology of supermodernity.* 1992 (J. Howe, Trans.). London: Verso.

Baelo-Allué, S. (2011). *Bret Easton Ellis's controversial fiction: Writing between high and low culture.* London: Continuum.

Benton, L. M. (1995). Will the real/reel Los Angeles please stand up? *Urban Geography, 16*(2), 144–164.

Buscall, J. (1999). Pretty vacant: MTV and de-individualization in Bret Easton Ellis's *Less Than Zero.* In S. Tanskanen, & B. Wårvik (Eds.), *Proceedings from the 7th Nordic Conference on English Studies* (223–231). Turku: University of Finland.

Colby, G. (2011). *Bret Easton Ellis: Underwriting the contemporary.* London: Palgrave.

Ellis, B. E. (1986). *Less than zero.* London: Picador.

Ellis, B. E. (1991). *American psycho.* London: Picador.

Ellis, B. E. (1995). *The informers.* London: Picador.

Ellis, B. E. (1998). *Glamorama.* London: Picador.

Ellis, B. E. (2010). *Imperial bedrooms.* London: Picador.

Haggerty, K. D., & Ericson, R. V. (2000). The surveillant assemblage. *The British Journal of Sociology, 51*(4), 605–622.

Lennon, J. R. (2010). Via 'Bret' via Bret. *London Review of Books, 32*(12), 21–22.

Lutton, A. (2012). East is (not) East: The strange authorial psychogeography of Bret Easton Ellis. *49th Parallel, 28,* n. pag.

Lutton, A. (2013). I just want to go home: Bret Easton Ellis's *Glamorama* and disturbed American transnationalism. In R. Davis (Ed.), *The transnationalism of American culture: Literature, film, and music* (pp. 121–138). London: Routledge.

Lyon, D. (2007). *Surveillance studies: An overview.* Cambridge: Polity Press.

Lyon, D., Haggerty, K. D., & Ball, K. (Eds.). (2012). *Routledge handbook of surveillance studies.* London: Routledge.

McQuire, S. (2008). *The media city: Media, architecture, and urban space.* London: Sage.

Murphet, J. (2001). *Literature and Race in Los Angeles.* Cambridge: Cambridge University Press.

Murphet, J. (2002). *Bret Easton Ellis's American Psycho: A reader's guide.* New York and London: Continuum.

Vincent-Arnaud, N. (2006). Cartographie du vide: les "non-lieux" de l'espace américain dans *The Informers* de Bret Easton Ellis. *Anglophonia: French Journal of English Studies, 19,* 107–116.

AUTHOR BIOGRAPHY

Alison Lutton is a Lecturer in English at Somerville College, University of Oxford. She teaches modern and contemporary literature and critical theory, and is currently preparing a monograph considering the relationship between authorship and literary value in the contemporary USA. Her other research interests include urban space and the creative process in American fiction, and social media and author and reader identities.

CHAPTER 8

Participating in '1984': The Surveillance of Sousveillance from *White Noise* to Right Now

Caleb Andrew Milligan

SURVEILLING THE SCENE

The debate seems lost: we *are* being watched. Our culture is a surveillance culture, for better and worse. The way to overcome that, a new generation has cheered, is through sousveillance: now we will do the filming and watching. It is not as easy, however, to just beat the watchers by watching. The gazes then only pile up, and the panopticon finds a new media distribution strategy. In this chapter, I make this argument by reading Don DeLillo's 1984 novel *White Noise* alongside the emerging technologies of that pivotal year to establish what I call the "surveillance of sousveillance" through tracing how those technologies have developed into our present cultural moment. I focus specifically on Jack's desire to create an "avatar," a character extension beyond himself that can reach the media network as a recognizable subject—from sousveillance viewer to surveillance victim—in Part II, "The Airborne Toxic Event." Surveillance—being constantly monitored by "eye-in-the-sky" technology—today is now reciprocally complicated by what Steve Mann has termed sousveillance—monitoring

C.A. Milligan (✉)
University of Florida, Gainesville, USA
e-mail: camilligan@ufl.edu

© The Author(s) 2017
S. Flynn and A. Mackay (eds.), *Spaces of Surveillance*,
DOI 10.1007/978-3-319-49085-4_8

back through "eyes-on-the-ground" technology. Where DeLillo's novel comes in theoretically handy, then, is its reflection of the media-saturated society in the year he completed writing it: 1984. The plot trajectory and thematic resonance of *White Noise* are both influenced by, and in reaction to, the permeating effects of digital technology. Though the book has been exhaustively analysed as a seminal text of television-age literature, DeLillo, as an author, is always in conversation with media, so I suggest we update that conversation to dialogue with new developments in network media. The novel's status as a text amid 1984's technological zeitgeist renders it, within my argument, DeLillo's prophetic text about the way we digitally manifest our extended identities into surveillance culture. 1984 was the year of the Apple Macintosh computer's debut (and the famous Super Bowl commercial advertising its arrival), as well as the interim between the Atari's demise and the Nintendo Entertainment System's North American release. Looking back, in this year when new media was on everyone's mind, Don DeLillo's *White Noise* stands out in its ability to predict the fallout of what happens when it is no longer just "them watching us," but now "us watching each other" in an interactive media panopticon of circulating content.

Surveillance is getting much bigger as it gets smaller. Mann et al. (2003, p. 334) explain that the Panopticon worked best in "pre-industrial 'door-to-door' communities," thus with the industrial revolution and the expanding of cities came a shift in the structure of surveillance. And with that shift, the emergence of "neo-panopticons": "Since that comparatively ancient time, surveillance techniques have increasingly become embedded in technology. Where people once watched people with their naked eyes, computer-aided machines now do remote monitoring of behaviour" (Mann et al. 2003, p. 335). The panopticon effect is digitally extended, as those recorded "do not have direct visual and aural contact with those who are observing" (Mann et al. 2003, p. 335) But the "computer-aided machines" have changed considerably since Mann et al. discussed them in 2003. Ten years later, Edward Snowden changed the conversation surrounding surveillance forever with his whistle-blowing revelations about NSA wiretapping and monitoring of American mobile phone records, a practice he has since revealed other countries are similarly guilty of. We now carry these neo-panopticons with us, as a majority of people in the developed world cannot get by without a smartphone.

Sousveillance supposedly offers a way to take this technology back. Steve Mann is considered the progenitor of the term and pioneer of the practice.

Mann et al. elucidate that since "sur" is French for "over," then "We call this...'sousveillance' from the French words for 'sous' (below) and 'veiller' to watch" (Mann et al. 2003, p. 332). To continue sampling Mann's terminology, he (Mann et al. 2003, p. 333) considers sousveillance a form of "reflectionism" in which the technologies used for surveillance are re-appropriated by the surveilled to film themselves, or even those surveilling them: "[it] holds up the mirror and asks the question: 'Do you like what you see?'" In our Web-enhanced society, not only must it be filmed, it must be circulated for social and, in Mann's original conception, political purposes. He proposes utopically:

> Digital technology can build on personal computing to make individuals feel more self-empowered at home, in the community, at school and at work. Mobile, personal, and wearable computing devices allow people to take the personal computing revolution with them. Sousveilling individuals now can invert an organization's gaze. (Mann et al. 2003, p. 336)

It is, however, the circulation of the sousveillant effort, I suggest, that compromises how radical Mann had it in mind to be. Sousveillance is rendered into web content, something to be seen as the consumer's form of entertainment. As surveillance technology reaches out further into realms of media, sousveillance ends up effecting, instead, an inverse reaching out, filming to be filmed. The surveillance of sousveillance undermines the reclaiming of technology for digital liberation as everyone remains entrenched in a media-entertainment-construct of the neo-panopticon.

1984 AND BEYOND

The shift from surveillance to sousveillance can be seen from *Nineteen Eighty-Four* to "1984." When George Orwell wrote his famous novel in 1948, he foresaw a society in constant deference to Big Brother, where the Ministry of Truth told lies and the televisions watched the viewers. It was a future of total and constant surveillance, in which the government is always a step ahead of Winston Smith's plot to push back against their dystopian control. Defeated and brainwashed, Smith joins in and admits that he loves Big Brother. There are few novels more crushing in their inescapable visions of totalitarian power. At the time, it threatened a future not that far away in which all privacy is past and the present is complete surveillance. Therefore, when 1984 actually came around, everyone could breathe sighs of relief that Orwell's vision had not come to pass. In fact, as I outlined earlier, tracing 1984's technological advancements to their contemporary developments, implicates that year as the beginning of a proto-sousveillance culture that encouraged interaction with organizational technologies tooled into the personal. The Apple Macintosh kicked off the personal computer revolution to come; suddenly this high technology was an interactive part of the home. But before the product, there was the pitch. The year in technology truly began with the most famous Super Bowl commercial to ever air, Apple's "1984" advertisement. More short film than commercial, this ad, directed by Ridley Scott, features a woman in bright red and white contrasted against a cold and grey future who hurls a hammer at, and shatters, the screen on which a "Big Brother" like authority captivates a uniformly dreary crowd of sad men with shaved heads. As the evocative scene concludes, Apple ends the commercial with an enticing tagline that states, 'On January 24th, Apple Computer will introduce Macintosh. And you'll see why 1984 won't be like "1984."' Using Orwell's dystopian classic as its context, Apple implicitly compares then-reigning IBM's lock on the computer market to a fate as dreadful as Big Brother's ideological dominance. This analogy makes Apple that hammer wielding woman. The Macintosh is boldly touted as the free-thinking individual's solution that will save society from a fate worse than a faceless dystopia. In the computing world of 1984, Apple made itself the unique hero against big institutional odds: To escape the surveillance nightmare of Big Brother Business, buy an Apple Macintosh.

Contributing his own proto-sousveillance prescience, DeLillo's *White Noise* contains similar thematic conflict between the interactive individual and the surveilled crowd. Just two years after the novel's publication, Tom LeClair published the first critical analysis of it, in which he famously calls DeLillo a "'systems novelist'…who analyse[s] the effects of institutions on the individual" (Osteen 1998, p. xii). Huddled under the umbrella term 'institution' could also be the panopticised public, which *White Noise* interacts with in the catastrophe of "The Airborne Toxic Event." Stacey Olster's (2008, p. 82) take on the text posits that, "The characters in *White Noise* can only locate themselves collectively within the crowd and by way of those places that facilitate congregation." With Jack in mind, however, fading into the examined crowd goes against his search for character throughout the novel. Rather than identify himself "collectively," Jack references himself against the crowd in order to become a distinct node in the system, for a system is necessary for identity formation, just as a "story" is needed for a character to emerge. His inverse surveillance is to be watchable rather than just merely watched. Jack would rather play the heroine with the hammer than be a part of the blank and grey crowd, recognizing that her red and white clothing needs that dreary palate to stand out. DeLillo's novels often feature what Susana S. Martins (2005, p. 90) calls "meditations on how individuals are forged within systems of language and ideology." We see in this novel that Jack plays a persona bigger than himself to reach beyond his individuality into the network. What his game entails, however, is an attempt to become a character, not quite a quest for realized identity. It is out of the crowd that Jack aims to escape surveilled self and attempt sousveilled extension. For within surveillance culture, Jack equates this technology that he cannot understand with death, hoping that his sousveillant attempts to connect to new media will let him "live."

In DeLillo's novel, new media and death often go hand in hand, both contributing to the *White Noise* of the title. When the SIMUVAC technician testing Jack for Nyodene D. exposure informs him that he is "generating big numbers" according to the computer, he frets and asks questions to hide his anxiety (DeLillo 1998, p. 140). The answers do not really help. Ultimately Jack finds out he is "the sum total of [his] data [and] No man escapes that," leaving him feeling "like a stranger in [his] own dying" (DeLillo 1998, pp. 141–142). Leonard Wilcox (1991) argues that this fear of technology directly correlates with the fear of death. According to his analysis (Wilcox 1991, p. 353), "the symbolic mediations of contemporary society deprive the individual of an intimate relation with death, with the result that society is haunted by the fear of mortality." Jack's "death" is rendered into computer code, indecipherable to him. He is being watched and recorded into the "massive data-base tally" (DeLillo 1998, p. 141). Therefore, the fear of death mediated into information may be seen as a vulnerability to surveillance. Jack then aims to invert this surveilled self through sousveillance: to become a character that does not just accept the code, but contributes to it. It is by inhabiting a role that one may approach death, through Mann's reflectionism. What is reflected, though, is not the thing itself. Martin Heidegger (2008, pp. 281–282), in *Being and Time*, offers that we can only understand death through the death of Others, which means never completely, for "the dying of Others is not something which we experience in a genuine sense; at most we are always just 'there alongside.'" We do not fully understand death then; more specifically, we do not want to understand it. Heidegger (2008) takes the phrase, "one dies," and then explicates that stubbornness. Through his idea of the "they," meaning everyone else but oneself, Heidegger (2008, p. 297) suggests "that what gets reached, as it were, by death is the 'they.'" One can say, "one dies," because everyone thinks '"in no case is it I myself,' for this one is the 'nobody'" (Heidegger 2008, p. 297). He says then that the "they" can even convince one that this person who will one day die is not oneself. One takes on a part and convinces oneself, "The person left to confront death is not me." The "they", however, are everyone else watching, constituting the surveillance of sousveillance. Being watchable through filmed acts of sousveillance only expands the neo-panopticon, which reaches farther than ever as phones are unwitting tools of surveillance and laptops, tablets, and smartphone hardware keep us within quick access of the Web wherever we are. When Google asks to access a user's location, the ability to invert that level of surveillance becomes even more difficult through interaction and contribution. We now seem to allow it, wanting Heidegger's "they" to be our audience. "Death has entered," says Jack, and he may be right (DeLillo 1998, p. 141): the surveillance of sousveillance has us even more connected but just as vulnerable.

In trying to escape his surveilled self, Jack's character play could be considered a game: the poignant medium through which one can "die" so much and yet live. For Jack and his game, the fact that virtually all digital games offer a player the chance for multiple lives is crucial. When one plays a game, he or she can perform clumsily and die quickly with the first life, and then that same player can perform well and complete a certain level with the next: both lives (and deaths) belong to the same character and yet do not. Jack, through his character, then may be said to be sampling lives by skirting close to death. These characters are not the actual Jack, but his extensions, the way he attempts to conceptualize death; because he never authentically faces death and, in fact, cannot without dying is exactly why he remains afraid of it. Fear of death, as fear of surveillance, here allows us to read DeLillo's media-saturated narrative as a cultural examination of lives perpetually at the mercy of screens. Before gaming was even a ubiquitous medium, DeLillo suggests through the character of Jack how we may engage those screens rather than just accept them. The interactive medium of gaming offers a proto-sousveillant way to approach death through playing it. The passivity of surveillance resets to the interactivity of sousveillant representation. New game.

The concept of becoming a separate sousveillant character reaches ideological completion in this new game, a media pastime that 1984's America oddly enough thought little of. Atari had just plummeted under the video game crash of 1983, because "no one," in the large economic sense, was playing it. Even though the Nintendo Entertainment System was exponentially gaining a reputation in Japan, it would not reach North America until the fall of 1985. Computer games mostly carried the year, but in a niche market. Therefore, this pivotal year is important for its omission of what is now a booming culturally legitimate medium. DeLillo has never written about video games; even a mention of the machines is nowhere to be found in *White Noise*. His text is, nevertheless, certainly prophetic of how we may conceptualize the way we become a character through gameplay as both first person player and third person character. DeLillo's first novel *Americana* predicts this provocative concept, albeit regarding television, when protagonist David Bell rehearses lines with an actor playing his father, who reads, "[Television] moves [man] from first person consciousness to third person. In this country there is a universal third person, the man we all want to be" (DeLillo 1989, pp. 270–271). A game, as a media technology with increased screen interaction, similarly affords the chance that "entering the third person singular might possibly be fulfilled" (DeLillo 1989, p. 271). The most telling part of these recited lines is the conclusion, the chance to enter the third-person singular. That third-person singular is nevertheless caught inside the pre-programmed plots of rule and mechanic bound story and gameplay, a visual representation of sousveillant play still trapped within a larger surveillance structure. Even the advent of online gaming and its freedom still promises that surveillance of sousveillance, all the players their own main characters, and everyone watching everyone.

To best conceptualize this representational reach into the network, I turn to Gregory Ulmer's (2011) avatar theory. Jack, as if he were online in real life, creates a character to adapt to his evolving situations in ways that project self beyond self. Not just the narrator of this novel as book, he actually narrates himself within the plot as well. Jack's narrative play over his character resembles a kind of real-life writing, but his persona projection, furthermore, evokes Ulmer's concept of avatar. In our own Internet age, Ulmer (2011) writes of the self-extending practice of becoming one's avatar, to inhabit one's online identity as something not "oneself" but another character entirely, for "Avatar is not mimetic of one's ego, but a probe beyond one's ownness." Explaining that, "The term avatar in

Sanskrit literally means descent," Ulmer (2011) discusses how the analogy comes from the times Krishna came down to earth and took on embodiment in Hindu mythology. Our own "descent" then is becoming our online selves, like Jack, the narrator, narrating his characterized extension. This play is a pastime born from what Ulmer (2011) calls the emerging language apparatus: "electracy." It is learning to communicate in the media of our increasingly digital culture, a step away from strictly print literacy to electronic fluency and production. To connect the literate to the electrate, he (Ulmer 2011) claims that, "Playing one's avatar is to electracy what writing an essay is to literacy." In his embrace of the digital apparatus, he (Ulmer 2011) encourages that, "You need to meet avatar, that part of you inhabiting cyberspace." If this is the case, then Jack's "character," a more literate construct, may be understood as "avatar" given the 1984 milieu discussed earlier. Within Ulmer's electracy, the technologies we interact with are tied to an apparatus that includes individual identity, so he suggests we employ avatar to understand where this emerging language apparatus is taking us and how we may intervene and take some control over it. I argue that if we analyze Ulmer's version of literacy as a more passive apparatus in relation to the interactivity of electracy, then we may likewise see surveillance as the passive overview against sousveillance as the gameplay on the ground. Meeting one's avatar, in light of that conception, should mean, more cynically, to reach out to be watchable rather than just to be watched; perhaps even to be a playable character than just played.

The relationship of avatar as described by Ulmer suggests that when one plays any game, he or she both is the character and is not. There is a sequence of identity extension in which one is oneself, one plays through the character, and one is the character. This relationship gets even more direct once we can design our own avatars rather than simply playing with pre-built characters, which is increasingly standard practice in certain console games and nearly all MMORPGs (Massively Multiplayer Online Role Playing Games). As innovator of Hitler Studies, Jack feels he must enlarge his persona (and waistline) to match his own academic import; in other words, he must become what a professor looks like. Per his chancellor's advice, he aims to "'grow out' into Hitler" (DeLillo 1998, p. 17). This embellished image grooming is what turns Jack into "the false character that follows the name around" (DeLillo 1998, p. 17). His dark glasses and robe even resemble mask and cape, making him seem like some super-professor and foreshadowing the hero that Jack wants to be when the Airborne Toxic Event cripples the town of Blacksmith. Jack's mission to bodily become a character entails playing his own life, designing his avatar of professor through his own sousveilled body to make it worthy of surveillance.

In times like these, however, as predicted by DeLillo, embodying a character is taken beyond gaming analogies bound by graphical interface, directly into players' bodies themselves. For example, the social video platform Twitch turns the players playing into DeLillo's "third person singular"—not just the characters they play—by fostering a network for anyone to upload streaming videos of their "Twitch plays," live footage of their gameplay for other players to watch. Through a vast community of Twitch "channels" (echoing the language of television at the heart of *White Noise*), players engage in sousveillance by filming themselves play their favourite games and, on average, exceptionally well, because the most talented players get the most interaction on their Twitch channels. Therefore, the community space of Twitch still rewards the extension into avatar, a character beyond just oneself, even when the character is oneself on camera. In fact, some Twitch players even make their livings off ad revenue associated with the views they receive. The surveillance of their sousveillance, then, is built into Twitch's platform as its intent is to create a shared community of everyone watching everyone (a neo-panopticon disguised in promises of fun and connection). Beyond merely real game players, advancements in augmented reality gaming, such as *Pokémon Go* are turning the real world into game space, just like we turn ourselves into sousveillant characters of the surveillance culture. While players "catch" virtual Pokémon in actual spaces via smartphone interfaces, the game highlights, even more clearly, the surveillance of sousveillance as the vast amount of data collecting permissions it requires players to grant has become the content of many news articles' cautions. These developments in gaming technology and culture continue to shift the proto-sousveillant medium of gaming into an actually sousveillant—and even surveillant—reflection of our implicated culture. As Jack Gladney learns in the face of the Airborne Toxic Event, becoming and playing a character is not a safe escape from the real anymore.

The Decentred Centre of the Sousveillance Disaster

In Part II of DeLillo's novel, "The Airborne Toxic Event," a black cloud of spilled Nyodene Derivative disrupts Jack's illusion of privileged safety and sends him clinging to the illusion of an Other. The illusion of the Other that I refer to directly correlates to the desire to be watchable. Just as I previously discussed how Jack references himself against the examined crowd, I claim that as we vie for this viewability, we separate ourselves from those we are privileged to watch. When that model will no longer last, we must extend ourselves through sousveillance into the right to be circulated as content. We then fall victim to the surveillance of sousveillance. As we see Jack shift from sousveillance viewer to surveillance victim within this state of emergency, we should notice that Jack's vulnerability to the television screen, as discussed by scholars like Martins (2005) and Wilcox (1991), nicely predicates what DeLillo can likewise say to networked windows. For it is within any screen, in fact, that this mediated Other can be found. As an example, when Jack and his children see Babette on television earlier, nothing short of a rediscovery takes place:

> The face on the screen was Babette's...I'd seen her just an hour ago, eating eggs,
>
> but her appearance on the screen made me think of her as some distant figure
>
> from the past, some ex-wife and absentee mother, a walker in the mists of the
>
> dead... It was but wasn't her... I tried to tell myself it was only television
>
> whatever that was, however it worked—and not some journey out of life or death,
>
> not some mysterious separation. (DeLillo 1998, pp. 104–105)

Jack here muses that Babette seems simultaneously more real and more mythical all at once. It is as if Jack's wife truly exists all over again for he and his family—not just as the Babette they thought they knew, but as that character in the television set. Wilcox (1991, pp. 346–347) says this is what happens in a world where "images, signs, and codes engulf objective reality; signs become more real than reality and stand in for the world they erase." The erased world gives way to the surveillance of everyone watching each other through screens, even those right alongside them in a disaster like this one that shakes Blacksmith.

Within the perceived world of mediated reality is a perceived need for this "Other." Jack as disaster viewer needs something to view, or more poignantly, needs a way to view himself. It is through this mediation that Jack compartmentalizes apocalypse and renders himself perceivably untouchable: bad things only happen to other people. Jack Gladney believes this lie wholeheartedly every time he assures his son Heinrich the cloud "won't come this way" and states so matter-of-factly:

> I'm a college professor. Did you ever see a college professor rowing a boat down his own street in one of those TV floods? We live in a neat and pleasant town near a college with a quaint name. These things don't happen in places like Blacksmith. (DeLillo 1998, p. 114)

He feels protected by the character he creates. This arrogance on Jack's part is comparable to the many people in our current culture who feel similarly unaffected by the pervasive surveillance permeating their own lives. It should take less than an "Airborne Toxic Event" to jar them into realizing that they cannot afford Jack's smugness. It is this realization on Jack's part as the cloud comes closer "this way" that puts his illusion in danger of disillusionment, as he puts off the idea that those who perceive their Others can likewise become someone else's Other. If people continue to think they are not being watched—just watching, then this false truism can be considered an example of Jacques Derrida's (1993) "centre." In *Writing and Difference*, Derrida (1993, p. 278) explains, "The function of this centre was…to orient, balance, and organize the structure—one cannot in fact conceive of an unorganized structure." The structure at play here is this othering effect, and at its centre is the surveillance victim as Other, drawing upon Derrida's (1993, p. 278) notion that "classical thought concerning structure could say that the centre is, paradoxically, within the structure and outside it." The Other may be the centre, but it is outside the structure, which means, "The centre is not the centre" (Derrida 1993, p. 279). Yet this centre that is not the centre still cannot hold. When Jack hears the sound of sirens announcing the chemical spill, what Derrida (1993, p. 280) calls a "rupture" occurs, because "when the structurality of structure had to begin to be thought, that is to say repeated" this centre is now decentred. Once Jack Gladney's frame of reference is disrupted, his narrative identity shifts and he is forced to face the fact that he is vulnerable to disaster too. This upset then requires him to think about the structurality of the othering effect as structure. The outcome of this decentring then renders him the centre. Martins (2005, p. 105) expounds upon this jarring re-evaluation by clarifying, "The startling thing about television's citationality is that sometimes, what's happening on TV is also happening to you." Martins's concept of "citationality" circa 2017 suggests that what is happening in the surveillance network is also happening to everyone: we in fact do it to each other. Now that this proto-sousveillant version is happening to Jack, it all comes full circle when those othering are themselves othered.

Jack the disaster victim is left then with the decision to realize that the decentred subject can become a new centre. In the chaos of the airborne toxic event, what transpires here is a heightened fervour for this new role disproportionate to its actual involvement. Jack then steps into his new identification as the character of another's Other at first implicitly and then more explicitly. At first, he quells his excitement, maybe does not even recognize it. As Jack drives away from the changing wind of Nyodene D., he does not realize how much he longs for decentred stardom when he complains:

> I wanted them to pay attention to the toxic event. I wanted to be appreciated for my efforts in getting us to the parkway. I thought of telling them about the computer tally, the time-factored death I carried in my chromosomes and blood. Self-pity oozed through my soul. I tried to relax and enjoy it. (DeLillo 1998, p. 159)

The disaster victim wants to play the hero now and be recognized for his daring escape. Unbeknownst to himself, he wants to show how much he has embraced his role as the Other in a dramatized disaster scene. In the surveillance of our sousveillance disaster, we similarly offer ourselves up as content: if someone is watching, then be worth watching.

The "recentered" centre then can be located in contemporary surveillance politics paradoxically through an infamous foil to Jack's disproportionate fervour to be the media hero. American police departments nationwide are under heavy scrutiny to equip their officers with body cameras after several highly covered police shootings of black civilians turned racial political discourse in the U.S. toxic. Operating under the assumption that everyone acts differently when they know they are being filmed, body cameras are seen as a transparent remedy to the inscrutability of police conduct in the vein of Mann's "watching the watchmen" reflectionism. But what they ironically undermine is the opacity of American police officers' hero narrative. If anyone questions the police, they are seen as doubting American heroes. Unlike Jack's sousveillant attempt to become the surveilled hero, police officers equipped with body cameras then must revert to ordinary model citizens. For without this forced accountability, these "American heroes" have had to be ratted out by citizen sousveillance. Black Americans have turned to smartphone technology for reflectionism against police brutality, such as the phone camera that caught Walter Scott's shooting on video and even Facebook Live, which immediately broadcast the aftermath of Philando Castile's murder. Therefore, as police officers are pressured to film themselves through wearable technology and citizens film them with their handheld versions, the "*white* noise" of these racially charged tragedies gets cut through by footage—for a moment. The title of DeLillo's novel remains apt, for what contributes to the white noise of media saturation is the surveillance of sousveillance. These socially justified uses of sousveillant practices have already become more content, one web search away from the watchable being watched. Yet still we try to answer surveillance with sousveillance just as Jack insists to be recognized by the only mediated gaze that matters: everyone else's.

CONCLUSION

Here in the blueprint of *White Noise* is where I argue that we similarly aim to interact with our media technologies through sousveillance in order to push back against the passivity of surveillance. We would rather be watchable than just be watched. Therefore, we contribute to the network in order to be a part of that network, instead of just a pawn. Nevertheless, the heroic disaster victim is still caught in the disaster, just as sousveillant connections end up only contributing to the surveillance of everyone watching each other. Many of Jack's fellow townspeople must similarly want to become their own heroes as well, so what was unique was likely never unique. As we circulate ourselves and our sousveillant attempts to invert surveillant gazes, we only contribute to the content that everyone else is too. The surveillance of sousveillance takes over, and we end up as disappointed as Jack and the rest of Blacksmith when the disaster was not even that bad: "a little weary, glutted in an insubstantial way, as after a junk food spree" (DeLillo 1998, p. 160). The centre is decentred as everyone both "watches you" and requests, "watch me." We are all the Others wanting to be viewed and viewing, therefore no one is. Forced to accept then that the surveillance of sousveillance is here to stay, not only was DeLillo right in 1984, but Orwell was onto something in *Nineteen Eighty-Four*: we all know we're being watched, so we give each other something to watch. By analysing Don DeLillo's *White Noise* alongside the emergent technologies of the year he wrote it, I argue that we see now how mediated we are than ever before through the ways we watch one another. The neo-panopticon is here; we helped it solidify its very power.

REFERENCES

DeLillo, D. (1989). *Americana*. New York: Penguin.
DeLillo, D. (1998). *White noise*. New York: Penguin.
Derrida, J. (1993). *Writing and difference* (A. Bass, Trans.). Chicago: University of Chicago.
Heidegger, M. (2008). *Being and time* (J. Macquarrie, & E. Robinson, Trans.). New York: HarperPerennial/Modern Thought.
Mac History. (1984). Video file. Retrieved October 23, 2013, from https://www.youtube.com/watch?v=VtvjbmoDx-I.
Mann, S., Nolan, J., & Wellman, B. (2003). Sousveillance: Inventing and using wearable computing devices for data collection in surveillance environments. *Surveillance and Society, 1*(3), 331–355.
Martins, S. (2005). *White noise* and everyday technologies. *American Studies, 46*(1), 87–113.
Olster, S. (2008). *White noise*. In J. Duvall (Ed.), *The Cambridge companion to Don DeLillo*. New York: Cambridge University.
Orwell, G. (2003). *Nineteen eighty-four*. New York: New American Library.
Osteen, M. (1998). Introduction. In M. Osteen (Ed.), *White noise: Text and criticism*. New York: Penguin.
Ulmer, G. (2011). Avatar emergency. *Digital Humanities Quarterly, 5*(3). http://www.digitalhumanities.org/dhq/vol/5/3/000100/000100.html.
Wilcox, L. (1991). Baudrillard, DeLillo's *white noise*, and the end of heroic narrative. *Contemporary Literature, 32*(3), 346–365.

AUTHOR BIOGRAPHY

Caleb Andrew Milligan is a Ph.D. student at the University of Florida in the Department of English. He currently researches in venues that combine literature and media studies, working with concepts of medium-specific analysis. His research includes print culture, electronic literature, film and game studies. He teaches undergraduate courses on close reading print and digital media.

CHAPTER 9

Surveillance in Post-Postmodern American Fiction: Dave Eggers's *The Circle*, Jonathan Franzen's *Purity* and Gary Shteyngart's *Super Sad True Love Story*

Virginia Pignagnoli

DIGITAL TECHNOLOGY AND SURVEILLANCE SOCIETIES

The Circle tells the story of Mae Holland, a woman in her early twenties who moves to Silicon Valley to take up her dream job in the world's most important high-tech corporation, the Circle. The Circle is a relatively young company that in less than six years has managed to buy out Facebook, Twitter and Google and create a platform called "TruYou," which combines in just one account all of our social media profiles, payment systems, passwords, email accounts (p. 21). In plain Orwellian fashion, the Circle supposedly works to make the world a better place and adopts slogans, such as "secrets are lies," "sharing is caring," "privacy is theft" (p. 303).[1] Similar to the "epic environmental dystopianism" of much twenty-first century fiction (Boxall 2013, p. 217), the dichotomy

V. Pignagnoli (✉)
University of Turin, Turin, Italy
e-mail: virginia.pignagnoli@unito.it

© The Author(s) 2017
S. Flynn and A. Mackay (eds.), *Spaces of Surveillance*,
DOI 10.1007/978-3-319-49085-4_9

utopia/dystopia couldn't be less unequivocal: the technological progress aimed at creating a utopian "imaginary ideal society that dreams of a world in which the social, political and economic problems of the real present have been solved" (Booker 2005, p. 127) creates instead a dystopia: "an imagined world in which the dream has become a nightmare" (ibid.). Such a nightmare is caused by the various digital innovations sold by the Circle and branded as essential for our well-being and for the progress of the human race. Digital technologies play a crucial role in the growth of a dystopian society ruled by surveillance systems. As remarked by David Lyon, "surveillance occurs in the most high-tech ways and at the pinnacles of power but depends on the humdrum, mundane communications and exchanges that we all make using online media and communication devices" (2015, p. vii).[2] Together with our growing engagement with online media and communication devices comes surveillance systems and loss of privacy. However, it seems that the utopian ideology that accompanied the digital revolution hasn't been supplanted yet. Not so much within the Internet pioneers themselves—Jaron Lanier, to mention one for all, has published widely against the ideology that "promotes radical freedom on the surface of the web" (2011, p. 3)—, but within the main Internet companies whose mantra is often to provide technological progress to make the world a better place. For instance, Mark Zuckerberg, Facebook's founder and CEO, in more than an occasion stated that Facebook's mission is to make the world more open and transparent, not for his company to become more profitable and powerful, but in society's interests (van Dijck 2013, p. 15). Reflecting this dynamic, *The Circle* transforms America into a totalitarian surveillance state with everyone's blessing. Echoing Mark Zuckerberg's words, the Circle's mantra invites its users to be more *transparent*, but transparency is just the sugarcoated synonym for surveillance: by providing the technology that allows people to record and broadcast every moment of their life, the Circle encourages the creation of a surveillance state, where politicians will commence, one after the other, to record and broadcast their lives in the name of democracy. It is this technology that, by keeping everyone happy like *Brave New World*'s soma, will lead to a world with no personal freedom.

Dystopian Spaces

If we define space as the environment in which characters move and live (Buchholz and Jahn 2005; Ryan 2014), within the narrative space of *Super Sad True Love Story*, constant surveillance is already a reality.[3] Everyone must wear a mobile phone (now called äppärät) "pendant-style around the neck" (p. 76), which projects and rates instant data about its owner (health status, bank account). Technological progress has led to a surveillance society, despite its utopian aspiration. Like in *The Circle*, many of these futuristic technologies are conveyed to appear not too unfamiliar. The characters interact through *GlobalTeens*, a Facebook-like social network where everyone in the world is traceable and that instead of verbal language promotes communication through images. In this dystopian New York, people are constantly looking at their äppärät, live streaming their lives, ranking other people and tracking them. When the äppäräts stop connecting because of a civil war, Lenny Abramov, the protagonist, writes in his diary: "We are all bored out of our fucking minds. My hands are *itching* for connection" (p. 268; italics added), underlining a physical bond with the digital device. But whereas Mae Holland becomes complicit in the creation of the Circle's dystopia, as she is not able to see the consequences of every little step "forward," Lenny tries hard to fit in with this nearly collapsed America obsessed by technology and youth, where instead of reading, people scan and privacy is a thing of the past. Lenny belongs to an elder generation and cannot avoid being nostalgic for a time when people would actually talk to each other instead of being constantly immersed in their äppäräts. He doesn't know how to use abbreviations properly; he has to learn how to "rate people quicker" (p. 65), and he will end up falling in love with a much younger girl, Eunice Park, who will eventually leave him for his boss (who is almost twice his age, but looks younger than him thanks to some pseudo-scientific technological progress). It is also thanks to Lenny's clumsiness that *Super Sad True Love Story* maintains the satirical tendency of dystopian fiction. In Shteyngart's novel, surveillance apparatuses become useful devices for tracking your girlfriend's location, "Fuckability" rankings are of crucial importance; there is a place called "Post-human services" that sells "Indefinite Life Extension" (it is where Lenny works); and every woman's aspiration is to work as a "Mediawhore."[4]

On the Novels' Ethical Dimensions

In the narrative space of Franzen's *Purity* the dystopian component is more circumscribed. *Purity*'s critique of digital technology, its utopian ideology and the surveillance techniques it enables, is diluted in a narrative of greater ambitions (*Purity* is a multilayered narrative sharply structured in seven long chapters employing an extra-diegetic narrator with different focalizations).[5] The ethical critique is conveyed mainly through a character, Andreas Wolf, a "famous Internet outlaw" from Germany (p. 16), who fought the communist regime back in Berlin and later invented the *Sunlight Project*, a sort of *WikiLeaks* enterprise committed to "honesty, truth, transparency, freedom" (p. 18). Unlike Julian Assange, however, Wolf can count on "universal admiration" (p. 57). Despite being a narcissistic archetype of evil (readers will soon discover that he killed a man at the end of the 1980s in East Berlin), in the fictional world he inhabits, he looks flawless: "Mr. Purity" (p. 61).[6] The ruling ideology of the *Sunlight Project* promotes freedom only on the surface (see Lanier above) and its motto is "secrecy is oppression and transparency freedom" (p. 17).

Franzen guides the ethical judgments of the readers by means of the juxtaposition of what the fictional characters perceive as the "pure" Andreas Wolf and what the readers know about him (all but pure).[7] The novel also allows us to draw a parallel between the claustrophobic atmosphere of the Soviet totalitarianism and the "new" apparatus Andreas Wolf is creating through the Internet and his Sunlight Project. This is conveyed through two mirroring chapters: *The Republic of Bad Taste* (Chap. 2) and *Moonglow Dairy* (Chap. 4). The former narrates of Wolf's life from his privileged childhood in the German Democratic Republic (his father being an undersecretary and his mother an English professor at the local University) to his troubled adolescence and further estrangement from his parents. The communist apparatus is portrayed as corrupt, shortsighted, but also ridiculous. Surveillance is everywhere, and it's the reason why Andreas wants to leave. Ironically, a similar surveillance society characterizes the Sunlight Project's headquarters in Bolivia described in the chapter titled *Moonglow Dairy*. Here, the readers follow Purity Taylor, aka Pip from Oakland, California, to the remote location of Los Volcanos where she will start an internship at Wolf's Project. Through Pip's eyes, we observe the people working at the Sunlight Project. Like those working at the Circle, they are young, with outstanding educational records, committed to make "*the world a better place*" (p. 243, original emphasis).

Operating once again through a deviation between the other characters' belated (or non-existing) awareness and the readers' knowledge, Andreas Wolf will unveil the obvious: "I've created a beehive of surveillance there" (p. 268), while Pip observes that "private electronic communication is impossible. The internal network was designed so that all chats and emails were viewable by anyone on the network" (pp. 282–283). Significantly, Wolf recreates in his Project the same totalitarianism, the same surveillance system he ridiculed while living in East Berlin. And just as we understand that the reasons for Andreas Wolf to fight the Republic are not morally just —he has very urgent personal reasons to do that, no less being a murderer and a pervert—so are the reasons behind his Sunlight Project. Everything he does is meant to hide his "real" self. Although his motto recites, "*Sunlight is the best disinfectant*" (p. 58, original emphasis), he only wants to hide his past by creating an image of himself as Mr. Purity. The message that what on the Internet appears as true, flawless, and transparent might as well be quite the opposite of that is quite straightforward.

The ethical dimension of *The Circle* is not only conveyed by the idea that technological progress leads to surveillance societies (see 1.1), but also through the creation of a sense of claustrophobia. The system *The Circle* is creating is a system ruled by sousveillance: the "democratization of broadcasting facilitated by mobile and perversive computing" (Mann and Ferenbok 2013, p. 19). In fact, to the Circle's ever-growing influence on every aspect of people's life corresponds a similar ever-growing control of the narrative space. Almost in its entirety, the novel takes place in the Circle's campus, a hyperbolical version of those of the various high-tech companies that populate Silicon Valley. If at first the campus, with its Calatrava fountain, picnic areas, tennis and volleyball courts (p. 1), kennels for the employees' dogs (p. 16), nightclubs, amphitheaters (p. 17), glass rooms, organic gardens, mini-golf areas, movie theaters, and bowling alleys (p. 29) is perceived as an idyllic, utopian place, soon enough it will become monopolizing. One of the major instabilities in the narrative progression indeed occurs after one of the few episodes in which the narrative space trespasses the limits of the Circle's campus: Mae ventures kayaking in the Pacific. After this incident, she will be brainwashed so that she will go "transparent" too, recording and broadcasting every moment of her life.

The role of the dystopian spaces described above (the collapsing New York, the uncanny perfection of Silicon Valley, the headquarters of Wolf's Project), with their panoptic style, is to serve the narratives' didactic purpose: the ethical dimension of the narratives is a critique of contemporary society that "implies (or asserts) the need for change" (Fitting cited in Tally 2014, p. 365). As we have seen, Franzen conveys this criticism through the parallelism between a totalitarianism and a journalist organization devoted to the online publishing of leaks and secret information labeled as truthful and transparent, but ruled by surveillance systems. In *The Circle*, the narrative space is entirely occupied by a high-tech company that becomes more powerful than any state or institution thanks to the information it holds. Shteyngart juxtaposes a clumsy Russian-American against the post-human inhabitants of a defeated New York. The narratives' warnings against the negative outcomes of societies more and more dependent on digital technologies is conveyed through the creation of spaces where this dependence has become a dystopian reality.

CLUELESS VOICES

According to rhetorical narratology, readers develop a specific kind of response in relation to the *mimetic* component of a narrative, which concerns "the characters as possible people and…the narrative world as like our own" (Phelan and Rabinowitz 2012, p. 7). Unlike the archetypical Wolf or the post-human Mae, Lenny's descriptions of New York are always nostalgic of the past: "the roaches, the somnolent minorities were gone, replaced by half-man, half wireless bohemians" (p. 80). Lenny's narration elicits a mimetic response whose concern is the future of humanity in a world ruled by digital hysteria. The mimetic response to *The Circle* is similar, although Eggers's novel employs archetypical characters, hinting to the genre of science fiction where ciphers or types are quite common (see Stockwell 2005, pp. 518–520). In charge of the Circle are the Three Wise Men: Ty, the young, Zuckerberg-style, creative mind; Bailey, the "happy and earnest" public face of the Circle (p. 24); and Stenton, the ruthless CEO who monetized Ty's utopia, having spotted right away the connection between "work and politics, and between politics and control" (p. 484). Mae and the other employees are puppet-like characters who express no interest in deep thought. Their dialogues, their tweets (here called "zings"), and their actions are extremely trivial, guided by an utopian desire to make the world a better place through digital technology. Mirroring the fact that "surveillance today comes not in the shape of a centralized and threatening state, but as manifold 'little brother' who do not affect us so much as citizens, but as consumers" (Kammerer 2012, p. 101), when the Circle starts selling cheap and easy-to-hide cameras, called "SeeChange," everyone wholeheartedly believes that they are invented to reduce the crime rate and advance democracy against totalitarian regimes. This also reflects Lyon's remark above and the idea that the responsibility of new technologies is, to some extent, also our own responsibility.

Unlike in *Super Sad True Love Story* where Lenny communicates a sense of nostalgia for a past without digital technology, in *The Circle* character-character narration is designed to communicate a sense of general unawareness (Lai-Tze Fan similarly refers to this as depthlessness [2016]). In the world ruled by the Circle's technologies, almost no one is able to recognize the danger of each single step forward. The kind of general unawareness that today allows private commercial companies to "hold potentially more information on their customers than any state institution" (Kammerer 2012, p. 101) is the same that in Eggers's novel will lead to a state of ubiquitous surveillance (politicians will be constantly spied on, chips will be implanted into children to prevent sexual abuse) while everyone is dazed by the consumer appeal of the various products. The presence of characters like Mae Holland—who don't need Orwell's *Newspeak* to be deprived of any form of critical thinking—is not accidental; it elicits the readers' mimetic responses to question these characters as versions of ourselves: are they/we human, post-human, cyborgs?[8]

Franzen employs a similar narrative strategy. The young people employed at the Sunlight Project are quite stereotypical characters, who uncritically trust technologies and, more importantly, those who own them. Pip, however, is not as dehumanized as Mae. After an initial infatuation with both the Project and his creator, she will eventually discover that it was not accidental that the most famous and generally admired person in the world emailed her personally with an invitation to Bolivia. A revelation to which will follow Andreas's confession: "I hate the Internet … I'm not doing this job because I still believe in it. It's all about me now. It's *my* identity" (p. 275). Despite this revelation and Andreas's confession, however, no one in the novel understands the danger of the Sunlight Project as Andreas himself does. And this danger is again expressed in terms of techno-consumerism: "like the old politburos, the new politburo styled itself as the enemy of the elite and the friend of the masses, dedicated to *giving consumers what they wanted*" (p. 449), says Wolf's voice through free indirect discourse.

Like Eggers, Franzen conveys his communicative purpose by relying on the readers' ability to negatively judge the lack of critical understanding in these characters, and the lack of values in their voices.[9] This mirroring, shallower version of ourselves in the near-future elicits a mimetic response that undoubtedly reinforces the narratives' ethical purpose. On the other hand, these clueless voices may also estrange the readers: flattened, unthinking characters are also predictable, and a predictable narration can make the reading experience less rewarding. Shteyngart avoids this possible side-effect by combining the naïveté of *Super Sad True Love Story*'s main character with an ironic mode. Lenny may be clueless as far as new technologies are concerned, but not because he is not able to see the faults in them, rather because he is clearly not interested:

> I really needed to figure out what this LIBOR thing was and why it was falling by fifty-seven basis points. But, honestly, how little I cared about all these difficult economic details! How desperately I wanted to forsake these facts, to open a smelly old book or to go down on a pretty young girl instead. Why couldn't I have been born to a better world? (Shteyngart 2010, p. 79)

Moreover, the narrative space concerning this apocalyptic digitization is not all-absorbing as in *The Circle*'s, but more fluidly blended with Lenny's and Eunice's adventures, their love story, their work careers, their friendships and relationships with their families. As opposed to a *defamiliarizing* effect common to dystopian fiction, in *Super Sad True Love Story* there is a plurality of voices, which conversely allows a *familiarizing* effect. Finally, unlike *The Circle* and *Purity*, the narrator is not omniscient. In fact, the narrating voice is mainly Lenny's, through his journal's entries, sometimes interrupted by Eunice's message exchange with her friends and family through GlobalTeens (the Facebook-like social network). Eunice, who at first seems like any other (future) teenager who cares only about appearances, will eventually become a brave supporter of a new Occupy movement at Tompkins Park. There is no post-human lack of moral values surfacing from Lenny's and Eunice's voices, allowing for a narrative whose communicative purpose is less didactic, less guided by an overly-overt authorial message against digital technologies.

POST-POSTMODERN NOVELS AND THE DIGITAL AGE: ON THE ETHICS OF THE AUTHORIAL PRESENCE ON THE INTERNET

In the first part of this essay we have seen how surveillance is represented in contemporary literature, the use of claustrophobic spaces, the dystopia resulting from ubiquitous social media, and the deceiving appearances of do-gooder organizations. Here, I would like to add a further layer of analysis; namely, I want to bring into the discussion the way the authors of these novels are, or are not, involved with the digital world they so openly criticize in their fictional works. This discourse on the authorial presence on the Internet is particularly significant for these post-postmodern novels dealing with surveillance because, as we will see, it is linked with their strong ethical commitment. But it also catches, I believe, an important aspect of contemporary literature that deserves more attention.

The idea that post-postmodernist fiction is supplanting postmodernism is relatively new, but critics have started to investigate some of its defining features and the term is becoming more and more commonly used when referring to twenty-first century literary narrative. Since its appearance, post-postmodern fiction has come to denote both a temporal shift—with either the new millennium or 9/11 as its starting date—and a shift in aesthetic values and literary strategies. For instance, Robert McLaughlin has described it as a response "both to the perceived exhaustion of literary postmodernism and to the growing dominance of television in popular culture" (2012, p. 12). Brian McHale, via Nealon (2012, ix), highlights the idea that post-postmodern symptoms involve an intensification and mutation within postmodernism (2015, p. 177), while Nicoline Timmer has argued that the post-postmodern novel signals a "turn to the human," with its focus on "what it means to be human today," on empathy and human interaction (2010, p. 361). Timmer's idea can be linked with the transition from irony to sincerity, a shift generally and famously associated with David Foster Wallace and his suggestion that irony lost its rhetorical power for narrative communication once a major medium device, the television, started employing it profusely. Perhaps the approach that has dealt the most with the new digital landscape is Alan Kirby's digimodernism (2009). Kirby mainly explores movies, games and TV shows, but his critical work has the merit of highlighting the relationship between digital media and more traditional narrative forms. Dorothee Birke and Birte Christ (2013), instead, started "mapping the field" for investigations on the connection between paratext and digitized narrative. Finally, in some of my recent work (Pignagnoli 2016b), I have drawn on Genette's concept of paratext too, bringing to the fore the twofold nature of Genette's term: composed of peritext and epitext. I suggest the new category of "digital epitexts" to comprise the digital elements officially produced or released by the author on authors' websites, blogs, social network sites, or as intermedial transpositions. Some of the functions of these digital epitexts include visual enhancement and reference to the print novel. Paul Dawson highlights the importance of nonfictional authorial voices that create a continuum with the narratorial commentary in the work of fiction (narrative statements whose ideological provenance cannot be attributed to a character) establishing reinforcing claims for an author's cultural capital (2013, p. 14). Among the examples given by Dawson there is Franzen's *The Corrections* (2001) and his essay "Why Bother?" (2002). According to Dawson, *The Corrections* is an "overt example of a novelist's deployment of omniscient narration as part of a broader project to reassert the authority of the novel in contemporary culture," while "Why Bother?" is the novel's nonfictional counterpart (Dawson 2013, p. 15). Also with *Purity* we can easily recognize a discursive continuum, as in several essays and interviews Franzen has commented upon digital technologies with remarks such as: "Twitter's and Facebook's latest models for making money still seem to me like one part pyramid scheme, one part wishful thinking, and one part repugnant panoptical surveillance" (2013). When in *Purity* we hear the omniscient narrator say, "If you substituted *networks* for *socialism*, you got the Internet. Its competing platforms were united in their ambition to define every term of your existence" (p. 448), a continuum in Franzen's fictional and nonfictional discourses is quite evident. Similarly, in a commencement speech he delivered at Kenyon College in 2011 (later adapted into an essay published in *The New York Times*), he criticized techno-consumerism and claimed that a person defined by a desperation to be liked (on social networks) is a "person without integrity, without a center...a narcissist—a person who can't tolerate the tarnishing of his or her self-image that not being liked represents." A person, thus, like Andreas Wolf, the "master dissembler" (p. 281). When in *Purity* the fictional narrative voice expresses through FID Andreas Wolf's thoughts, we may as well hear the whole Franzen-weltanschauung about the Internet:

> There were a lot of could-be Snowdens inside the New Regime, employees with access to the algorithms that Facebook used to monetize its users'

privacy and Twitter to manipulate memes that were supposedly self-generating. But smart people were actually more terrified of the New Regime than of what the regime had persuaded less-smart people to be afraid of, the NSA, the CIA—it was straight from the totalitarian playbook, disavowing your own methods of terror by imputing them to your enemy and presenting yourself as the only defense against them—and most of the could-be Snowdens kept their mouths shut. (p. 450)

So, why would all this matter? Because these nonfictional discourses reinforce the ethical dimension of *Purity*. To Franzen's fictional critique of digital technology corresponds a nonfictional critique expressed through many essays and interviews. And part of Franzen's "cultural capital" is also his non-active presence on the Internet and social media, an absence that reinforces his fictional and nonfictional critiques (for Franzen to share his thoughts on Twitter or his pictures on Instagram could seem a little inconsistent).

While Dawson mainly refers to authors' essays, interviews, and writers' festivals as where we can hear the extrafictional authorial discourses, what I wish to highlight is a further nonfictional authorial voice emerging from their presence on the Internet. Many authors, indeed, share opinions and personal life experiences through social media or support their fictional narratives with extra-fictional material on websites or blogs. Shteyngart, for instance, is among those contemporary writers who embrace the new communication channels offered by digital media and often shares pictures on Instagram. His "voiced" presence on the Internet opposes Franzen's, as well as Eggers's, but it doesn't necessarily weaken the ethical message conveyed through *Super Sad True Love Story*, it being, as we have seen, less guided by didacticism.[10] Moreover, with *Super Sad True Love Story*, Shteyngart participates in the debate around the cultural status of contemporary literature by making explicit references to a future world where all the physical libraries have closed and no one is interested in reading anymore. To everyone, books smell "like wet socks" (p. 35), but not for Lenny, who is still able to "REALLY, REALLY READ instead of scan" (p. 145). It emerges a clear anxiety about the "decline of book culture" (Dawson 2013, p. 247) through the character-narrator's voice: "No one but me still cares about you. But I'm going to keep you with me forever. And one day I'll make you important again" (p. 50). And again, "I've spent an entire week without reading any books or talking about them too loudly," Lenny writes in his diary (p. 76), "I'm learning to worship my new äppärät's screen, the colourful pulsating mosaic of it, the fact that it knows every last stinking detail about the world, whereas my books only know the minds of their authors" (p. 76). Similarly, in *The Circle* and *Purity*, the presence of clueless characters who despise reading in favor of a flattened digital communication reveals a meta-discourse about the cultural authority of books: in a future dominated by communication through digital media, the kind of deep communication that reading allows is denied.

Eggers's fictional discourse in *The Circle,* conversely, benefits from his "unvoiced" presence on the Internet. The ethics of *The Circle* resonates with the warning that "our feeling that we're entitled to know anything we want about anyone we want" is the greatest threat to our freedom today, as he claimed in a recent interview (*The Telegraph*). By creating a dystopian world inhabited by clueless characters, Eggers explores, as he puts it, the implications of technology "for our sense of humanity and balance" (*McSweeney's*). The omniscient narrative voice that in *The Circle* reports the creation of a surveillance society thanks to the oblivious support of dull post-humans like Mae Holland can be considered an invocation of Eggers's authorial voice when he says that "over the last 20 years, it's been interesting to see how little resistance there is to the merging of our organic selves and the devices that we attach to ourselves to enhance our capabilities" (*The Telegraph*).

And, similar to Franzen, the ethical values at work in Eggers's novel are the same ethical values that restrain him from having an active presence on social media: the narrative voices that in *The Circle* say things like, "tyrants can no longer hide. There needs to be, and will be, documentation and accountability. ALL THAT HAPPENS MUST BE KNOWN" (67) are also invocations of Eggers's "unvoiced" presence on the Internet. In other words, to avoid the use of digital resources, as Franzen and Eggers do, or to employ them, as Shteyngart does, reveals a commitment towards certain ethical values; ethical values that may, or may not, be in conflict with those at work in the authorial fictional discourses, influencing the narrative communication. This is why to exclude these digital nonfictional discourses from a narrative analysis means to exclude a significant rhetorical exchange between an author and her readers.[11] Finally, by referring to Franzen's, Eggers's and Shteyngart's presence on the Internet as "voiced" or "unvoiced," I do not aim at defining a clear-cut categorization. As Jonathan Safran Foer recently acknowledged, "it's not an either/or—being 'anti-technology' is perhaps the only thing more foolish than being unquestioningly 'pro-technology'—but a question of balance that our lives hang upon" (2013).

CONCLUSION

Surveillance is "spreading and intensifying" (Murakami Wood 2013, p. 317) and contemporary American literature responds to our new networked culture with narratives that have a strong ethical dimension. By showing how the increasing power of digital technology may have brought about societies where transparency is just a synonym for surveillance, *The Circle*, *Super Sad True Love Story*, and *Purity* offer a quite explicit ethical warning. As Mann and Ferenbok have remarked, "with media, we become sousveillance-enabled individuals able to contribute to a broader social responsibility of undersight" (p. 32). Whether we will "contribute to under sight or become co-opted into complicit drones in corporate and larger government surveillance networks, remains to be seen" (ibid.). The post-postmodern novels analyzed here, depict near-future worlds where utopian aspirations evolve (or they are likely to evolve) into Orwellian dystopias. Freedom is only superficial and characters are semi-human lacking of critical thinking.

As I showed, these novels' critique of contemporary society is reinforced (*The Circle*, *Purity*) or challenged (*Super Sad True Love Story*) by the authors' voiced, or unvoiced, digital presence on the same media that is criticized at the fictional level. This is significant with regard to an ethical reading of such novels, but also for a more general discussion of post-postmodernism.[12] To conclude with an open question, Glenn Greenwald recently asked, "Will the digital age usher in the individual liberation and political freedoms that the internet is uniquely capable of unleashing? Or will it bring about a system of omnipresent monitoring and control beyond the dream of even the greatest tyrant of the past?" (2014, p. 6). The novels analyzed here seem to lean towards the second scenario but it will be interesting to observe how this literary trend will respond, in the future, to possible changes in surveillance practices following evolutions in computing and new technologies.

NOTES

1. As a reminder, the famous slogans in *Nineteen Eighty-Four* are: war is peace/freedom is slavery/ignorance is strength (1961, p. 4).
2. See also "Surveillance, Power and Everyday Life" (Lyon 2007).
3. For a recent discussion of the concept of space in narrative studies see Joshua Parker's "Conceptions of Place, Space and Narrative: Past, Present and Future" (2016).
4. For a thorough discussion of genre in *Super Sad True Love Story* see also Lai-Tze Fan (2016).
5. The only exception is the fifth chapter, *[lelo9n8a0rd]*, the memoir of Tom Aberant, the character narrator.
6. Franzen portrays the digital world as a place where appearances are extremely deceiving, e.g. despite the "multitude of haters on the Internet," for Mae, it is difficult to find "hostile comments about Wolf" (p. 58).
7. Readers refer to rhetorical readers: "those member of the actual audience who seek to discern the authorial design behind a narrative's various ways of communicating" (Phelan 2014, p. 52).
8. Mann and Ferenbok state that wearable computing "can turn every wearer into an information production factory—a surveillance-capable post-cyborgian human" (p. 29). For a concise but effective discussion of posthumanism see F. Ferrando (2013).
9. With "values in their voices" I am hinting at Phelan's definition of voice as the "synthesis of style (diction and syntax), tone (a speaker's attitude toward an utterance), and values (ideological and ethical)" (2014, p. 49).
10. Shteyngart's digital presence may also be a way for him to adopt a sharing strategy consistent with many contemporary authors' attempts to find new modes of sincerity (authors who adopt this sharing strategy are writers who share their creative process to offer their readers information they may need to understand the communicative purpose(s) of their narratives). See also Pignagnoli (2016b).
11. This claim would also require a discussion on authorship and the implied/real author that I cannot undertake here. For an overview of the concept of the implied author and its development in narrative theory see Dan Shen (2013).
12. And to the rhetorical approach to narrative if we argue, as I do, for the digital presence of authors (voiced/unvoiced) as something worth including into our rhetorical analysis. See Phelan (2014) for the rhetorical model of narrative communication and Pignagnoli (2016a, b) for a preliminary discussion of digital resources in the narrative communication model.

References

Birke, D., & Christ, B. (2013). Paratext and digitized darrative: Mapping the field. *NARRATIVE, 21*(1), 65–87.

Booker, M. K. (2005). Dystopian fiction. In D. Herman et al. (Eds.), *Routledge encyclopedia of narrative theory* (pp. 127–128). London: Routledge.

Boxall, P. (2013). *Twenty-first-century fiction: A critical introduction.* New York: Cambridge University Press.

Buchholz, S., & Jahn, M. (2005). Space. In D. Herman et al. (Eds.), *Routledge encyclopedia of narrative theory* (pp. 551–554). London: Routledge.

Dawson, P. (2013). *The return of the omniscient narrator: Authorship and authority in twenty-first century fiction.* Columbus: Ohio State University Press.

Eggers, D. (2013). *The circle.* London: Hamish Hamilton.

Fan, L. (2016). The digital intensification of postmodern poetics. In T. Lanzendörfer (Ed.), *The poetics of genre in the contemporary novel* (pp. 35–55).

Ferrando, F. (2013). Posthumanism, transhumanism, antihumanism, metahumanism, and new materialisms: Differences and relations. *Existenz, 8*(2), 26–32.

Fitting, P. (2010). Utopia, dystopia, and science fiction. In G. Claeys (Ed.), *The Cambridge companion to Utopian literature* (pp. 135–53). Cambridge: Cambridge University Press.

Foer, J. S. (2013). How not to be alone. *The New York Times.* Retrieved March 6, 2016, form http://www.nytimes.com/2013/06/09/opinion/sunday/how-not-to-be-alone.html?_r=0.

Franzen, J. (2011). Liking is for cowards. Go for what hurts. *The New York Times.* Retrieved March 6, 2016, from http://www.nytimes.com/2011/05/29/opinion/29franzen.html.

Franzen, J. (2013). What's wrong with the modern world. *The guardian.* Retrieved September 6, 2013.

Franzen, J. (2015). *Purity.* New York: Farrar, Straus and Giroux.

Greenwald, G. (2014). *No place to hide: Edward Snowden, the NSA, and the U.S. Surveillance State.* New York: Metropolitan Books.

Kammerer, D. (2012). Surveillance in literature, film and television. In K. Ball et al. (Eds.), *Handbook of surveillance studies* (pp. 99–106). London: Routledge.

Lyon, D. (2007). Surveillance, power, and everyday life. In R. Mansell et al. (Eds.), *The Oxford hand book of information and communication technologies* (pp. 449–472). Oxford and New York: Oxford University Press.

Lyon, D. (2015). *Surveillance after Snowden.* Cambridge: Polity Press.

Mann, S., & Ferenbok, J. (2013). New media and the power politics of Sousveillance in a surveillance-dominated world. *Surveillance and Society, 11*(1/2), 18–34.

McHale, B. (2015). *The Cambridge introduction to postmodernism.* New York: Cambridge UP.

McLaughlin, R. (2012). Post-postmodernism. In J. Bray, A. Gibbons, & B. McHale (Eds.), *The Routledge companion to experimental literature* (pp. 212–223). New York: Routledge.

McSweeney's. (2013). A brief Q&A with Dave Eggers about his new novel. *The Circle*. Retrieved March 6, 2016, from http://www.mcsweeneys.net/articles/a-brief-q-a-with-dave-eggers-about-his-new-novel-the-circle.

Murakami Wood, D. (2013). What is global surveillance? Towards a relational political economy of the global surveillant assemblage. *Geoforum, 49*, 317–326.

Nealon, J. (2012). *Post-postmodernism: Or, the cultural logic of just-in-time capitalism*. Stanford, CA: Stanford University Press.

Orwell, G. ([1949] 1961). *Nineteen eighty-four*. New York: Signet Classics.

Parker, J. (2016). Conceptions of place, space and narrative: Past, present and future. *Amsterdam International Electronic Journal for Cultural Narratology* (AJCN), (7 & 8), 74–101 (Autumn 2012/2014).

Phelan, J. (2014). Voice, tone, and the rhetoric of narrative communication. *Language and Literature, 23*(I), 49–60.

Phelan, J., & Rabinowitz P. (2012). *Narrative theory: Core concepts and critical debates*. With D. Herman, B. Richardson, & R. Warhol. Columbus: Ohio State University Press.

Pignagnoli, V. (2016a). Sincerity, sharing and authorial discourses on the fiction/nonfiction distinction: The case of Dave Eggers's you shall know our velocity. In T. Lanzendörfer (Ed.), *The poetics of genre in the contemporary novel* (pp. 97–111).

Pignagnoli, V. (2016b). Paratextual interferences: Patterns and reconfigurations for literary narrative in the digital age. *Amsterdam International Electronic Journal for Cultural Narratology* (AJCN), (7 & 8), 102–119 (Autumn 2012/2014).

Ryan, M. L. (2014). Space. In P. Hühn, Peter et al. (Eds.), *The living handbook of narratology*. Hamburg University. Retrieved March 6, 2016, from http://www.lhn.unihamburg.de/article/space.

Shen, D. (2013). Implied author, authorial audience, and context: Form and history in Neo-Aristotelian rhetorical theory. *Narrative, 21*(2), 140–158.

Shteyngart, G. (2010). *Super sad true love story*. London:Granta.

Stockwell, P. (2005). Science fiction. In D. Herman et al. (Eds.), *Routledge encyclopedia of narrative theory* (pp. 518–520). London: Routledge.

Tally, Robert T. (2014). Lost in grand central: Dystopia and transgression in Neil Gaiman's "American gods." In Grubisic, Baxter, & Lee (Eds.), *Blast, corrupt, dimantle, erase: Contemporary North American dystopian literature* (pp. 357–371). Waterloo: Wilfrid Laurier University Press.

The Telegraph. (2013). Dave Eggers interview. The Telegraph. Retrieved March 6, 2016, from http://www.telegraph.co.uk/culture/books/authorinterviews/10356543/Dave-Eggers-interview.html.

Timmer, N. (2010). *Do you feel it too? The post-postmodern syndrome in American fiction at the turn of the millennium*. New York: Rodopi.
van Dijck, J. (2013). *The culture of connectivity: A critical history of social media*. New York: Oxford University Press.

Author Biography

Virginia Pignagnoli is a post-doctoral fellow at the University of Turin (Italy). Her latest publications include "Narrative Theory and the Brief and Wondrous Life of Post-Postmodern Fiction" (*Poetics Today* 2018), "Paratextual Interferences: Patterns and Reconfigurations for Literary Narrative in the Digital Age" (*AJCN* 2016), and "Sincerity, Sharing and Authorial Discourses On the Fiction/Nonfiction Distinction: The Case of Dave Eggers's You Shall Know Our Velocity" (2016). She is currently completing her book-length study on contemporary fiction and digital paratexts.

CHAPTER 10

Citizen: Claudia Rankine, From the First to the Second Person

Jeffrey Clapp

Claudia Rankine's *Citizen: An American Lyric* (2014a) and its precursor *Don't Let Me Be Lonely: An American Lyric* (2004) have become two of the most galvanizing books of poetry published this century. Their impact is the result, in part, of their formal experimentation, including Rankine's dialogues with images and artworks, and her work's fluid blend of poetry and prose. Equally, however, these poems' power results from Rankine's return to two very familiar topoi: the rights and responsibilities of the citizen, and the form and meaning of the lyric. Despite the apparent distances between those two sets of ideas, Rankine has been considering contemporary citizenship through the lens of lyric form. As her lyric becomes an inquiry into what it now means to be a citizen, Rankine envisions how both lyric, and citizenship, are implicated with, and inflected by, contemporary practices of surveillance.

Don't Let Me Be Lonely and *Citizen* address two basic dimensions of surveillance experience: the responses to contemporary terrorism usually discussed under the rubric of "national security", and the hypersurveillance of black citizens in the United States, particularly by the police. In both of these areas, Rankine arrives at the conclusion that will become the focus of

J. Clapp (✉)
Department of Literature and Cultural Studies,
Education University of Hong Kong, Hong Kong, China
e-mail: jmclapp@eduhk.hk

© The Author(s) 2017
S. Flynn and A. Mackay (eds.), *Spaces of Surveillance*,
DOI 10.1007/978-3-319-49085-4_10

this essay's consideration of contemporary spaces of surveillance: the advent of the surveillance society has confused and challenged, in an essential way, the "visibility regime" (Brighenti 2010)—that is, the assemblage of rhetorics, practices, and spaces of visibility—that undergirds liberal democracy. From this point of view, Rankine has been writing about how surveillance, and particularly racialized surveillance, participates in what Wendy Brown and others have been describing as the "undoing of the demos" (Brown 2015) that comes with the imposition of neoliberal governmentalities, and particularly surveillance practices.

Rankine's work tends to confirm important surveillance studies contributions that articulate how, from the Middle Passage to the plantation, the Atlantic slave trade was as much a laboratory of modern surveillance practices as was Bentham's cylindrical prison (Parentl 2003; Murzuclf 2011; Browne 2015). So these poems certainly explore the real opposition between surveillant and democratic practices. But by directing her attention to citizenship in particular, Rankine also develops the idea of an underlying contiguity between surveillance and democracy. For Anthony Giddens:

> Movements oriented to the enlargement of democratic participation within the polity should be seen as always ... oriented toward redressing imbalances of power involved in surveillance ... There is a basic flaw, however, in the [idea] that the expansion of organizations inevitably supplants 'democracy' ... The intensification of surveillance [is] the condition of the emergence of tendencies and pressures toward democratic participation. (Giddens 1985 p. 314)

Democracy, that is, does not merely pre-exist the surveillance which threatens it; equally, the development of democracy results from surveillance itself, a conclusion equally important to both Tocqueville and Weber (Dandeker 1990). Moreover, surveillance studies scholars have recently been emphasizing that *citizenship* is inextricable from surveillance—it specifically requires identification, cataloguing, tracking, and the discrimination of classes of individuals (Lyon 2010). At the same time there has been a notable return to citizenship on the left, as, critical theorists like Jacques Rancière and Étienne Balibar have been arguing that it is primarily, or even exclusively, through the claim of citizenship that democracy can be renewed (Rancière 1999; Balibar 2015). From multiple points of view, therefore, and particularly in the case of citizenship, the possibility of even distinguishing the surveillant and the democratic seems to falter.[1] Forms of "representation", and even forms of "self-expression", can be equally well described in either idiom.

By locating her recent work directly within this problematic, Rankine contributes to our understanding of contemporary surveillance practices, especially insofar as they conform to what Rachel Hall has termed an increasingly ubiquitous "aesthetics of transparency" (Hall 2007, 2015). For Hall, citizens are now rendered, and make themselves, *transparent* across numerous domains—whereas those who remain *opaque* become subject to suspicion, detention, and violence. Hall's figuration might well be seen as a specification of Nicholas Mirzoeff's argument that in contemporary visuality, authority makes an increasingly exclusive claim to be "able to look"—against which he would assert for the people a counterpoised "right to look" (Mirzoeff 2011). But Rankine's work does not quite reach an equally instrumental formulation. Instead, for example, in *Don't Let Me Be Lonely* we witness Rankine at the airport, like everybody else, being asked to drink from her own bottle of water, and to take off her shoes —the now-familiar rituals of "security theater" (Schneier 2009) which we supposedly undergo to demonstrate that our bottles and shoes do not contain explosives, but which also inculcate the aesthetics of transparency across new and expanding domains.

Among those domains, as we will see, is the space of Rankine's poetry. That poetry explores how the aesthetics of transparency conceptually interferes with the visibility regime that has obtained in classical liberal democracy. Liberal democrats maintain, for example, that citizens have a right to representation, the right to be seen and known. And those who critique the state, the law, or even liberal democracy itself, generally accept this visibility regime, contending that most problems are failures of visibility: persons and groups are unjustly rendered invisible. But there is no self-evident correlation between the axis *visible/invisible* and the axis *transparent/opaque*. That which is transparent can be understood as visible or invisible; that which is opaque—the same. In fact, the aesthetics of transparency renders *invisibility* strangely proximate to *hypervisibility*, which is to say that it renders surveillance society strangely proximate to the society of the spectacle. An immediate consequence, as Rankine remarks in *Citizen*, is that "No amount of visibility will alter the ways in which one is perceived" (2014a, p. 24).

While Rankine's work now intervenes directly in this confused conjunction, she has been considering the visibility regime of American democracy since the first poem of her first book, published in 1994. In "American

Light" Rankine is already making the gestures that lead her to subtitle both *Don't Let Me Be Lonely* and *Citizen* "An American Lyric". The poem ana-tomises a partial citizenship, and it figures that partiality as a scarcely-perceptible shadow which is itself cast by the brilliantly lit, and brilliantly American, space of the public. The title of the book, *Nothing in Nature is Private*, similarly suggests the desire to step into the light: what sanction does nature give to our obsession with privacy? From her first book, Rankine calls into question the frequently reactionary insistence on privacy as the antidote to, or opposite of, surveillance (Nelson 2002; McGrath 2004). At this point there is an extended indented quotation from the poem "American Light". I do not know why it has been eliminated here.

What "American Light" does not yet call into question, however, is the rhetoric by which increased visibility serves as the rejoinder and response to a shadowed and partial civic existence. The train of thought which led Rankine to her present interrogation may have begun, however, with the attack on Rodney King by police in 1992.[2] In 2014, Rankine discussed the beating as crucial for her understanding of her position in American space:

> I know when Rodney King's jury came back and said that despite the video, the police had done nothing wrong, that was a moment for me. I literally burst into tears. I had this weird feeling walking around the streets of New York, that I didn't know who these people were … Because I think I always sort of believed in the justice system before that, even though I knew the history. I still felt that when you're not leaving it up to hearsay, when you have documentation, people will step up. And it didn't happen. That was really a crisis moment for me. You just feel like, okay, you need to start paying attention. It's the same line, from Rodney King to Michael Brown. It's a continuum. (Rankine 2014b)

The death of Michael Brown, shot by police in 2014, has become emblematic of the fact that black communities in the United States remain subject to a level of police surveillance and violence not manifestly distinct from that of the era of Emmett Till, of W.E.B. Du Bois, or of Frederick Douglass. Rankine's response to the King beating merits specific attention because it begins to reread visibility not as representation or recognition, but instead as a kind of null value, one which may well be powerless to defend victims of police violence, or to relieve Rankine's intensifying sense of vulnerability in spaces like "the streets of New York".[3]

Rankine's comment on King begins to lead toward the idea that visi-bility might be a kind of "trap". This need not immediately mean what

Foucault meant by it; there are many reasons to believe that the conceptual collapse of the surveillant into the panoptic occludes as much as it reveals (Mathiesen 1997; Haggerty and Ericson 2000; Wood 2007). For Rankine, visibility is a trap for reasons that have less to do with the institutional-ization of discipline and disciplines, and more to do with the canonical idea that visibility, recognition, and representation can serve as solutions to injustice and inequality. Rankine touches directly on the single most wide-ranging consequence of the increasing implication of surveillance society and democratic culture: their rivalry over what it means, in each of its registers, literal and metaphorical, to be visible.

Watching, being watched, and watching the watchers are key themes in *Don't Let Me Be Lonely.* Rankine's poetic eye joins that of the military and the media to peer everywhere, even into Saddam Hussein's "very red" mouth during the medical examination which followed his capture—the exact same moment which Rachel Hall has used to illustrate the "aesthetics of transparency" (Rankine 2004, p. 124; Hall 2007). But even as Rankine joins millions of others to peer intently into that mouth, she also writes, in an immediately political way, about the fear of being unseen. At this point there is an extended indented quotation from *Don't Let Me Be Lonely.* Where is it? Why has it been eliminated?

Not convinced of the interpretation, perhaps, but Rankine's point is driven home obliquely: the lack of interest in "convincing" other people of interpretations gestures toward a degradation of democracy—in the very context of democracy's core component, the election.

Where the husband does not 'lift his gaze' in *Don't Let Me Be Lonely,* in *Citizen* Rankine takes up many other ways in which a person's presence fails to result in that person's ethical and political recognition. The book repeatedly transitions from narrating experiences of *invisibility* to narrating experiences of *hypervisibility,* beginning with the former:

> You are twelve attending Sts. Philip and James School on White Plains Road and the girl sitting in the seat behind asks you to lean to the right during exams so she can copy what you have written … Sister Evelyn never figures out your arrangement perhaps because you never turn around to copy Mary Catherine's answers. Sister Evelyn must think these two girls think a lot alike or she cares less about cheating and more about humiliation or she never actually saw you sitting there. (Rankine 2014a, p. 5–6)

Citizen offers many more anecdotes of racialized humiliation like the one above, and Rankine insistently frames them as experiences of invisibility:

> Yes, and you want it to stop, you want the child pushed to the ground to be seen, to be helped to his feet, to be brushed off by the person that did not see him, has never seen him, has perhaps never seen anyone who is not a reflection of himself. (Rankine 2014a, p. 17)

Further:

> Apparently your own invisibility is the real problem causing her confusion (Rankine 2014a, p. 43).   .

Rankine's framing suggests that to be invisible is not really the same as to receive "black looks", those "epidermalizing" gazes which link skin colour to race, and link race to racist caricature (Hooks 1992; Browne 2015). Though she does sometimes write about that kind of gaze, Rankine is specifically focused on understanding the racializing gaze, not as discriminatory or derogatory, but as simply incapable of sight.

We can see this more fully by noting how Rankine figures racialized hypervisibility in a parallel way. In *Citizen*, Rankine writes at length about watching Serena Williams play tennis on television, and how the tennis court and tennis culture more generally, make manifest the dilemmas of racial visibility. While, on the one hand, "Serena and her big sister Venus brought to mind Zora Neale Hurston's 'I feel most colored when I am thrown against a sharp white background'" (Rankine 2014a, p. 25), Rankine also figures Serena's body as occluding sight itself:

> Though no one was saying anything explicitly about Serena's black body, you are not the only viewer who thought it was getting in the way of [the referee's] sight line. (Rankine 2014a, p. 27)

From the invisible body we move to the body that is opaque. It is worth emphasizing that Rankine figures Serena's situation through the ideas of the seen and the unseen, and particularly as that which is so visible that it prevents something else, namely the position of a tennis ball, from being seen. This is a curious and perhaps yet more complicated kind of blindness, but one that shares the same central idea as Rankine's other anecdotes: not now the racializing and discriminatory, but the visible and invisible as such.

Rankine extends this logic to the street policing which forms a central element of racialised surveillance in the contemporary United States:

> I left my client's house knowing I would be pulled over. I knew. I just knew. I opened my briefcase on the passenger seat, just so they could see … And you are not the guy and still you fit the description because there is only one guy who is always the guy fitting the description. (Rankine 2014a, p. 105–109)

Here Saddam Hussein's opened mouth has become an opened briefcase, and Serena's hypervisibility has become the same as that of the victim of the "driving while black" paradigm: even though this driver does not resemble anybody, when perceived, he is so intensely visible that his image blots out any particular person's description.

Rankine has described a literary response to this kind of immediate visibility. Because "To be a person of color in a racist culture is always to be addressable … So a writer of color may be fueled by a desire to exit that place of addressability" (Rankine and Loffreda 2015, p. 16). In her own work, Rankine says that she has moved away from "truth", through "transparency", to create a "recess":

> [Another poet suggested that *Don't Let Me Be Lonely*] was simple and direct in order to perform truth-telling. This reading of the style of the book surprised me because I worked hard for simplicity in order to allow for projection and open-endedness in the text, for a sort of blankness and transparency that would lose the specificity of 'the truth' … I am not interested in narrative, or truth, or truth to power, on a certain level; I am fascinated by affect, by positioning, and by intimacy … What happens when I stand close to you? What's your body going to do? What's my body going to do? On myriad levels, we are both going to fail, fail, fail each other and ourselves. The simplicity of the language is never to suggest truth, but to make transparent the failure. The linguistic failures are disappointing and excoriating, as you say, and the images don't exactly recoup or repair—they are a form of recess … (Rankine 2014c)

In *Don't Let Me Be Lonely*, Rankine arrives at a clear counterpoint with her earlier enunciation of "longing for the American light". In choosing "blankness", "transparency", and the idea of a "recess" to describe her writing, Rankine explicitly identifies an impasse within contemporary regimes of visibility. That impasse is not broken by identifying and denouncing invisibility. Instead, Rankine writes toward the aesthetics of transparency—even if that means abandoning the modes of speaking that are conventionally aligned with the claim to be "visible" or "represented", like "truth-telling" or "speaking truth to power". Thus the affective register of Rankine's new poems is "failure", and failure is itself a representation of how Rankine understands citizenship. In a refrain that recurs in *Citizen*—and in that book's only explicit discussion of citizenship—Rankine writes: "Yes, and this is how you are a citizen: Come on. Let it go. Move on" (2014a, p. 151).

What is to be "let go" is the affective burden imposed by the experiences of invisibility and hypervisibility that Rankine relates. So despite the fact that visibility would seem to be the necessary mediator by which an individual's presence would lead to recognition, visibility in Rankine does not have this function. Instead, Rankine's juxtaposition of invisibility with hypervisibility frames visibility as a locus of what Lauren Berlant has called "cruel optimism":

> 'Cruel optimism' names a relation of attachment to compromised conditions of possibility whose realization is discovered either to be impossible, sheer fantasy, or too possible, and toxic. What's cruel about these attachments, and not merely inconvenient or tragic, is that the subjects who have x in their lives might not well endure the loss of their object or scene of desire, even though its presence threatens their wellbeing. (Berlant 2010, p. 20)

Rankine has referenced Berlant's work in discussing *Citizen*:

> In *Cruel Optimism*, Berlant talks about things that we're invested in, despite the fact that they are not good for us and place us in a non-sovereign relationship to our own lives. And I thought, on a certain level, that thing that I am invested in that is hurting me would be this country [laughs]. (Rankine 2014b)

An *American* lyric. Berlant's description of the cruelly optimistic corresponds very closely to Rankine's articulation of contemporary visibility: on the one hand, for those rendered invisible, it is impossible to achieve; at the very same time, it is all too easy to become hypervisible. This double bind is reproduced at multiple scales in *Citizen*, not least in paradoxes, like "trying to dodge the buildup of erasure" or "fighting off the weight of nonexistence" (Rankine 2014a, p. 11, 139).

For Rankine, citizenship is the apotheosis of cruel optimism's exhausted endurance. In *Don't Let Me Be Lonely*, it's that "You think that voting won't make a difference" (Rankine 2004, p. 127). In *Citizen*, not only does citizenship amount merely to "mov[ing] on", but from the very first line of *Citizen*, there is a sense of weariness: "When you are alone and too tired even to turn on any of your devices ..." (Rankine 2014a, p. 1). Lauren Berlant has herself attempted to recuperate *Citizen's* gestures of exhaustion, noting that:

> *Citizen's* great phrase about your being 'too tired even to turn on any of your devices' ... is metapoetic but also implies that the maneuver of tone is one of your citizen-actions, a weapon for resisting defeat and depletion in the face of the supremacist ordinary. (Rankine 2014c)

Berlant seems to refer to an unembellished frankness she perceives in Rankine's style, a style that Berlant implies is without rhetorical "devices". But Rankine's sentence is metapoetic not only because of the play on the word "device", but also because in this very sentence Rankine specifically "turns on" not only "any" device, but what is in some ways the device of devices: apostrophe, in the form of Rankine's unusual use of the word "you".

As several of the citations above have suggested, much of *Citizen* is written "in the second person", a *you* who suffers both the imbecile and the cruel aspects of living in a racist culture. Of course, one way to read this unusual "point of view" would be to see in this "you" an attempt to "put the reader in the shoes of" somebody who is familiar with those experiences.[4] But as we have seen, Rankine does not frame herself as a teller of truths to those who do not know them. Rankine's "you" is not primarily a kind of direct address, but rather a use of, or even a reference to, apostrophe: "A figure of speech in which a thing, a place, an abstract quality, an idea, a dead or absent person, is addressed as if present and capable of understanding" (*Penguin Dictionary of Literary Terms*).

To read Rankine's "you" as apostrophe rather than interpellation is to honor her insistence that her recent work constitutes an American *lyric*. While the idea of lyric is a congeries of concepts and histories and involves many other elements than the apostrophic, the tradition of J.S. Mill, taken up by Northrop Frye in *Anatomy of Criticism*, seems to be directly upheld in Rankine's recent work:

> [In lyric,] to go back to Mill's aphorism ['eloquence is heard, poetry is overheard'] ... the poet, so to speak, turns his back on his listeners. (Frye 1957, pp. 249–550)

Frye goes on to define lyric in an even more pertinent manner: "a literary genre characterised by the assumed concealment of the audience from the poet"

(Frye 1957, p. 366). Many commentators have tended, perhaps following Paul de Man, to attribute a certain absoluteness to the disjunction that Mill and Frye have described as merely an obliqueness or turn, claiming that a poem can *never* address an audience (de Man 1984; Smith 2007). Conversely, others insist that the lyric is hardly spoken in solitude:

> Criticism tends to read acts of speaking to you as the doomed attempts to 'reconcile' alienated minds through a series of rhetorical strategies. That is to say, the apostrophic model is misleading, insofar as it has encouraged subjectivist interpretations of poetry … It is time, I think, that this long familiar notion that an I calls to a you, but cannot confirm anything beyond its own unanswered and unanswerable speech, is met with an anti-subjectivist, anti-Cartesian account of address. (Pollard 2012, p. 11)

For Rankine's lyric, however, the truth lies, as in her account of citizenship, somewhere between the solipsistic and what one might call the "dialogic" accounts—which is to say, directly within the classical account that gives us the lyric as a "turning away" and an "assumed concealment".[5] In fact *Citizen* all but explicitly writes through Mill's well-known account of the lyric:

> I they he she we you turn
>
> only to discover
>
> the encounter
>
> to be alien to this place.
>
> Wait.
>
> […]
>
> Overheard in the moonlight.
>
> Overcome in the moonlight. (Rankine 2014b, pp. 140–141)

The moonlight that plays across this poem, in which there is a turn away, an expectant pause, and the merest "overhearing," is the etiolated reprise of what Rankine offered in her first book's floodlit "American Light". For this reason, the new work may not be well-described as what Berlant terms a "weapon" (Rankine 2014c) on behalf of a citizen's action; instead, to quote Rankine herself, it seems that *Citizen* is invested in "a truce with the patience of a stethoscope" (2014a, p. 156).

Rankine's lyric "you" does not offer the creation of what she calls an "encounter". Like visibility and citizenship, the notion of a lyric encounter is another form of cruel optimism. Rather than striking through the poem to a direct prose style or to "direct address", Rankine's "you", and the designation "American lyric", both designate the oblique overhearing that constitutes a theory of the lyric, and a theory of the citizenship, that obtain within surveillance space. Rankine has said that she collected the anecdotes in *Citizen* not only from her own life, but also from friends and colleagues. So it is possible to conceive that she might have simply adapted them using the conventions of the memoir, in the first person. Or she might have hidden them behind fictional names, in the third person. But Rankine had already tried those strategies—in a certain sense they are the strategies of hypervisibility on the one hand, and invisibility on the other: "Tried rhyme, tried truth, tried epistolary untruth, tried and tried" (2014a, p. 71). Remaining, then, is a confrontation with the contemporary regime of the aesthetic of transparency on something like its own ground: a turn away, an overhearing in moonlight.

The image Rankine chose to feature on the cover of *Citizen* is a work of art, David Hammon's *In the Hood* (1993). The installation comprises simply the hood from a grey hoodie, affixed to a wall. Remediated by Rankine from its original context, it now inevitably refers to the 2012 death of Trayvon Martin, a young man wearing a hoodie, whose murder by a "neighborhood watch" volunteer resulted in many American celebrities and politicians donning hoodies in protest themselves. But what is a "hoodie", in the context of Rankine's work, and in the context of the aesthetics of transparency in which and through which she is writing? Given a certain cruel optimism about the stakes of contemporary visibility, the hoodie represents a compromise position, a truce. On the one hand, of course, it covers: it creates opacity and even a kind of "recess". On the other hand, via the aesthetics of transparency, that very opacity attracts attention and renders the wearer hypervisible.

The "recess", the "turn away", the "overheard", the "hoodie": *Don't Let Me Be Lonely* and *Citizen* trace what we might best understand as what Malcolm Bull has termed a "coming into hiding" (Bull 1999). For Bull, hiding is not becoming less knowable, less present; that would be invisibility, absence, silence. On the contrary:

[S]omething fully known going into hiding involves becoming less know-able ... so for something unknown, hiding involves becoming more knowable ... In other words, being hidden does not involve going into hiding, but coming into hiding. (Bull 1999, p. 26)

For the contemporary citizen, rendered all but transparent within spaces of surveillance, "Emancipation [can be] experienced as a coming into hiding" (Bull 1999, p. 255). From this point of view, what we are looking at in Rankine's delicate approach to the aesthetics of transparency is not a move "backwards" from a politics of visibility on the one hand, or from a poetics of the encounter on the other hand, but instead a movement that she would like us to regard as future oriented: if a failure, a failure forward.

The same obtains in Rankine's understanding of the experience of citizenship, which has been enjoying a kind of sustained theoretical renaissance. In 1992 Chantal Mouffe had already declared that:

A radicalization of the modern democratic tradition...can be achieved through an immanent critique, by employing the symbolic resources of that very tradition ... There is a consensus on the left that we should revive the idea of *citizenship*. (Mouffe 1992, p. 1, 3)

For Étienne Balibar, similarly, one does not hope to introduce ideas from outside the domain of democracy; instead one must work to constantly redefine citizenship, without any hope of producing another, superior concept (Balibar 2015, p. 124–125). What is notable about these theoretical figurations is their resemblance to the kind of "truce" described by Rankine, both in *Citizen* and in interviews. She is at once chastened and sanguine:

[I go] back to this idea of connection, community, and citizenship. You want to belong, you want to be here. In interactions with others you're constantly waiting to see that they recognize that you're a human being ... The truce is that. You forgive all of these moments because you're constantly waiting for the moment when you will be *seen*. As an equal. As just another person. As another *first* person. There's a letting go that comes with it. I don't know about forgiving, but it's an 'I'm still here'. And it's not just because I have nowhere else to go. It's because I believe in the possibility. I believe in the possibility of another way of being. Let's make other kinds of mistakes; let's be flawed differently. (Rankine 2014b)

CONCLUSION

Until the advent of the first person, the flaw Rankine has chosen is lyric apostrophe. She has designated at the level of genre the situation of the citizen under surveillance, and in particular, the situation of the highly visible who yet remain strangely invisible. Much of the response to Rankine's work has focused on how it speaks to the recent Black Lives Matter movement, and the way that Rankine explores the intimate and affective dimensions of living in a racist society that constantly trumpets its triumph over racism. But Rankine's work also points at a longer history and a wider scope of experiences in which, as she writes "Everything shaded everything darkened everything shadowed // is the stripped is the struck—" (Rankine 2014a, p. 146). That conjunction of paradoxes denotes the contemporary aesthetics of transparency that now appears to be emerging across multiple domains. It is Rankine's particular contribution to show how that aesthetics at once emerges from, and threatens, a democracy founded on the principle of equal citizenship, and it is her particular insight that lyric form can still conjoin, and perhaps even contain, these tensions.

NOTES

1. Or, putting the point more colloquially: 'the desire to watch and be watched is a more deeply rooted element of the liberal democratic impulse than we care to admit' (Pecora 2002, p. 345). For Pecora as for many others, the new technologies of visibility, from reality television to social media, are not external to democratic culture but of its essence (Pecora 2002).
2. It is not only Rankine who remembers the Rodney King beating as a crucial moment for understanding contemporary regimes of visibility. The incident is regularly recalled as the central example of 'counter-surveillance' or 'sousveillance': in short, the watching of the watchers (Mann et al. 2003; Monahan 2006; Browne 2015).
3. Writing about King in *Citizen*, Rankine compares the way that the images of Rodney King's beating, caught on video, 'trumped all other images'; whereas following Mark Duggan's death at the hands of police in the U.K., no images were recorded to contest the police narrative (Rankine 2014a, pp. 115–118). Though focused on the U.S., Rankine's exploration is a global one (Rankine 2014a).
4. *Guernica*: Talk to me about your decision to set many of these poems in the second person. *Claudia Rankine*: There were a number of things going on. Because some of the situations were mine and some belonged to other people, I didn't want to own them in the first person ... But that

was the least of it. The real issue was, the second person for me disallowed the reader from knowing immediately how to position themselves. I didn't want to race the individuals. (Rankine 2014b)

5. Because of the ongoing debate about the political valences of lyric, one critic attempts to evade the problem thus: "As its title suggests, Rankine's most recent book *Citizen: An American Lyric* (2014a) develops the writer's postlyric poetics." (Tucker 2015). I think Rankine's generic claim should be taken at face value (Tucker 2015; Rankine 2014a).

REFERENCES

Balibar, É. (2015). *Citizenship*, T. Scott-Railton, Trans., Cambridge: Polity.

Berlant, L. (2010). Cruel optimism, In M. Gregg & G. J. Seigworth (Eds.), *The affect theory reader*. Durham & London: Duke University Press.

Brighenti, A. M. (2010). Democracy and its visibilities. In K. D. Haggerty & M. Samatas (Eds.), *Surveillance and democracy* (pp. 51–68). Abingdon and New York: Routledge.

Brown, W. (2015). *Undoing the demos: Neoliberalism's stealth revolution*. Brooklyn: Zone Books.

Browne, S. (2015). *Dark matters: On the surveillance of blackness*. Durham and London: Duke University Press.

Bull, M. (1999). *Seeing things hidden: Apocalypse, vision, and totality*. London: Verso.

De Man, P. (1984). *The rhetoric of romanticism*. New York: Columbia University Press.

Dandeker, C. (1990). *Surveillance, power and modernity: bureaucracy and discipline from 1700 to the present day*. Cambridge: Polity.

Frye, N. (1957). [1990], *Anatomy of criticism*. Princeton & Oxford: Princeton University Press.

Giddens, A. (1985). *The nation-state and violence: Volume two of a contemporary critique of historical materialism*. Cambridge: Polity.

Hall, R. (2007). Of ziplock bags and black holes: The aesthetics of transparency in the war on terror. *The Communication Review, 10*, 319–346.

Hall, R. (2015). Terror and the female grotesque: Introducing full-body scanners to U.S. airports. In R. E. Dubrofsky & S. A. Magnet (Eds.), *Feminist surveillance studies* (pp. 127–149). Durham & London: Duke University Press.

Haggerty, K., & Ericson, R. (2000). The surveillant assemblage. *British Journal of Sociology, 51*(4), 605–622.

Hooks, B. (1992). *Black looks: Race and representation*. Boston: South End Press.

Lyon, D. (2010). Identification, surveillance, and democracy. In K. D. Haggerty & M. Samatas (Eds.), *Surveillance and democracy* (pp. 34–50). Abingdon & New York: Routledge.

Mann, S., Nolan, J., & Wellman, B. (2003). Sousveillance: Inventing and using wearable computing devices for data collection in surveillance environments. *Surveillance and Society, 1*(3), 331–355.

Mathiesen, T. (1997). The viewer society: Michel Foucault's "panopticon" revisited. *Theoretical Criminology, 1*(2), 215–234.

McGrath, J. E. (2004) *Loving big brother: performance, privacy, and surveillance space*. Routledge, London & New York.

Mirzoeff, N. (2011). *The right to look: A counterhistory of visuality*. Durham & London: Duke University Press.

Monahan, T. (2006). Counter-surveillance as political intervention? *Social Semiotics, 16*(4), 515–534.

Mouffe, C. (1992). Democratic politics today. In C. Mouffe (Ed.), *Dimensions of radical Democracy: Pluralism, citizenship, community* (pp. 1–16). London: Verson.

Nelson, D. (2002). *Pursuing privacy in cold war america*. New York: Columbia University Press.

Parenti, C. (2003). *The soft cage: surveillance in America from slavery to the war on terror*. New York: Basic Books.

Pecora, V. P. (2002). The culture of surveillance. *Qualitative Sociology, 25*(2), 345–358.

Pollard, N. (2012). *Speaking to you: Contemporary poetry and public address*. Oxford: Oxford University Press.

Rancière, J. (1999). *Disagreement: Politics and philosophy*, J. Rose, Trans. Minneapolis: University of Minnesota Press.

Rankine, C. (1994). *Nothing in nature is private*. Cleveland: Cleveland Poetry Center at Cleveland State University.

Rankine, C. (2004). *Don't let me be lonely: An American lyric*. St. Paul: Graywolf Press.

Rankine, C. (2014a). *Citizen: An American lyric*. Minneapolis: Graywolf Press.

Rankine, C. (2014b). Blackness as the second person. C. Rankine interviewed by M. Sharma, *Guernica* (17 November 2014). Retrieved January 20, 2016 from https://www.guernicamag.com/interviews/blackness-as-the-second-person/.

Rankine, C. (2014c). 'Claudia Rankine', C. Rankine interviewed by L. Berlant, *Bomb* (Fall 2014). Retrieved January 20, 2016 from http://bombmagazine.org/article/10096/claudia-rankine.

Rankine, C, & Loffreda, B. (2015). Introduction. In C. Rankine, B. Loffreda, & M. K. Cap (Eds.), *The racial imaginary: Writers on race in the life of the mind*. Albany: Fence Books.

Schneier, B. (2009). Beyond security theater. *New Internationalist*. Retrieved January 19, 2016 from https://www.schneier.com/essays/archives/2009/11/beyond_security_thea.html.

Smith, J. M. (2007). Apostrophe, or the lyric art of turning away. *Texas Studies in Literature and Language, 49*(4), 411–437.

Tucker, J. A. (2015). Waking up to the sound. *American Literary History*. Retrieved January 19, 2016 from https://alh.oxfordjournals.org/content/early/2015/06/11/alh.ajv033.full.

Wood, D. M. (2007). Beyond the panopticon? Foucault and surveillance studies. In J. Crampton & S. Elden (Eds.), *Space, knowledge and power: Foucault and geography* (pp. 245–263). Aldershot: Ashgate.

Author Biography

Jeffrey Clapp is an Assistant Professor at the Education University of Hong Kong working in the Department of Literature and Cultural Studies. He is currently writing a book on surveillance, democracy and life writing in contemporary US literature. In 2015, he co-edited a collection of essays entitled *Security and Hospitality in Literature and Culture: Modern and Contemporary Perspectives* (Routledge 2015).

PART III

States, Place and Bodies

Castrating Blackness: Surveillance, Profiling and Management in the Canadian Context

Sam Tecle, Tapo Chimbganda, Francesca D'Amico and Yafet Tewelde

"There's no difference between the North and South. There's just a difference in the way they castrate you."— (Baldwin et al. 1989, p. 45)

In *Dark Matters: On the Surveillance of Blackness* (2015), Simone Browne writes that surveillance is an essential part of black life (p. 6). Taking Blackness as "metaphor and lived materiality," Browne's study opens with her repeated attempts through the Freedom of Information Act (FOIA) to obtain CIA surveillance files on Frantz Fanon. The time period of Fanon's life in which Browne was interested was toward the end of his life (October 3-December 6 in 1961) when Fanon travelled to Washington, DC in attempts to prolong his life by receiving treatment for myeloid leukemia. Browne has faced challenges in obtaining any real information. According to the CIA and FBI, though gravely ill and incapacitated, Fanon still merited surveillance. Browne's work establishes a long history of surveillance of black individuals by the FBI, for example, James Baldwin, Ol'

S. Tecle (✉) · T. Chimbganda · F. D'Amico · Y. Tewelde
York University in Toronto, Toronto, Canada
e-mail: samtecle@yorku.ca

© The Author(s) 2017
S. Flynn and A. Mackay (eds.), *Spaces of Surveillance*,
DOI 10.1007/978-3-319-49085-4_11

Dirty Bastard of the Wu-Tang Clan, and Fanon. The FBI appears to have targeted Black radicals, artists, activists and intellectuals (Browne 2015, p. 3). Targeted individuals had, in common, a yearning for Black freedom and equality expressed through written word, art, music and other mediums. These expressions, according to the FBI and CIA necessitated surveillance. Browne's study establishes a critical and imperative understanding that at the heart of struggles for freedom and equality in America, the free Black body, even imagined, is taboo. Accordingly, we turn our attention to the practice of surveillance of Black lives in Canada. We examine the psychic effects that surveillance has wrought on individuals and communities of Black people. Constituted in and through antiblackness, we posit that surveillance, as framed by Browne, is a mechanism of castration, as framed in psychoanalytic theory (Chimbganda forthcoming) In this context, castration refers to the psychic oppression, management, and control of Black lives through policies and practices that cast Black people in the role of aliens rather than citizens within Canada.

In 1971, under then Prime Minister Pierre Elliot Trudeau (1968–1979), Canada implemented its Multiculturalism Policy. The policy was presented as a cultural mosaic in which diverse groups of people were empowered to flourish, something akin to "the American Dream." Canadian multicultural policy, recognized by the Canadian Charter of Rights and Freedoms in 1982, ensures that every Canadian receives equal recognition and treatment, and that the nation recognizes and protects Canada's multicultural heritage, aboriginal rights, and the rights of minorities to enjoy their distinctiveness in an egalitarian cultural landscape. The policy had four basic pillars: assisting cultural groups in retaining and fostering their identity; assisting cultural groups in overcoming barriers to their full participation in Canadian society; promoting creative exchange among all Canadian cultural groups; and assisting immigrants in acquiring at least one official language (English or French). Following that, Multiculturalism Policy was enshrined into the constitution when the Charter of Rights and Freedoms was patriated to Canada from Britain in 1982. This culminates with the passing of the Multiculturalism Act in 1988.[1]

We decry the depiction of multiculturalism as Canada's most exalted national integration policy and highlight that antiblackness, surveillance, and racial profiling serve to keep Black lives in a space of inferiority and oppression by way of racism. In doing so, we demonstrate the need for a novel understanding of what social, political, and economic castration can do to entire communities of Black people as an epidemic, not just in the United States of America, but in Canada too. The surveillance of Blackness facilitates racial profiling resulting in the continued casting of Blackness, both historically and ontologically, as a problem (DuBois 1904). The psychological suffering of Black lives under surveillance is not a factor of the past, but a real and present form of castration that has lasting effects. Within the discourse of Canadian multiculturalism, our framework offers a perspective on surveillance as a mechanism, which exposes the underlying ugliness of Multiculturalism as an oppressive act of governance over those deemed foreign.

By using and extending the concepts of surveillance, racial profiling and management, and applying them to sites of higher education, popular culture, and activism within Canada, we reveal that Blackness, and by extension Black people, within and under Canadian multicultural discourse, are persistently marked as problems to integration, unity, and national identity. As such, the state and its various institutional arms determine that Blackness requires constant surveillance, profiling, and management, which according to Browne maintains the status quo of antiblackness. Our conception uncovers the realities of the celebrated narrative of multiculturalism, as a mechanism used to separate one body of persons from others, eliciting explanations and demands that Black people justify their place within the national discourse. Black people are limited in their freedoms and national participation through profiling. The chapter will detail the ways in which the physical and psychic oppressive management of Black lives presents challenges for navigating the liminal paradoxical reality of national hopefulness and singular oppression.

In order to better understand how the state is invested in controlling the activities of Black people in Canada, we examine three particular case studies. The first details the experiences of Back male graduate students in higher education. It highlights the challenges of racial profiling faced by Black people as they attempt to navigate and achieve success in the ivory tower. The second case study focuses on the regulation of Black cultural expression in popular culture. Within the sphere of art and culture, Black expression must be managed and controlled in order to maintain a "Canadian national identity." Finally, the third case study extrapolates on state-controlled spying mechanisms. It exposes the shady justification of surveillance of Black lives as necessary for national security.

Black scholars and activists have highlighted the precarious positionality of Black life historically and at present. Essentially, the free Black body is taboo. Over time, the taboos have altered and consequently been represented differently; however, the underpinning ideology and discourse has crystallized. Examples of historically significant safeguards against taboos in the discourse of antiblackness include segregated schooling, and the criminalization of interracial marriages between Black and White people. It was taboo for Black people to sit in front of the bus, for Black people to have sexual or romantic relationships with White people, and it was taboo for Black people to own property (Austin 2013). Humanitarian notions of freedom, democracy, and equality have not eradicated the fundamental structure of the taboo. It has simply become more refined and more expertly disguised. Freud states, "For us the meaning of taboo branches off into two opposite directions. On the one hand it means to us sacred, consecrated: but on the other hand it means, uncanny, dangerous, forbidden, and unclean." (Freud and Strachey 1950) He explains that taboo restrictions are different from religious or moral prohibitions and cannot be traced to a commandment of a god, but actually impose their own prohibitions, and are differentiated from moral prohibitions by failing to be included in a system which declares abstinences in general to be necessary and gives reasons for this necessity. Taboo prohibitions lack all justification and are of unknown origin, which though incomprehensible to us, are taken as a matter of course by those who are under their dominance. Freud concludes that properly speaking, taboo includes only (a) the sacred (or unclean) character of persons or things, (b) the kind of prohibition which results from this character, and (c) the sanctity (or uncleanliness) which results from a violation of the prohibition (Freud and Strachey 2000).

Browne's gesture in *Dark Matters* (2015), of placing the field of surveillance studies in the long history of the transatlantic slave trade, is significant because it reveals "the surveillance of blackness as often unperceivable within the study of surveillance, all the while blackness being that nonnameable matter that matters the racialized disciplinary society" (Browne 2015, p. 6). She states that surveillance "locates blackness as a key site through which surveillance is practiced, narrated, and enacted" (Browne 2015, p. 9). Given the psychoanalytic understanding of the structure of taboos, when we examine the limitations and prohibitions that Black lives have experienced historically, they cannot be justified through moral, legal or even scientific reasoning. Understanding the underpinning values in the practice of surveillance and control is crucial to understanding why approaching the subject from a legalistic, moral or rights based position cannot yield the desired outcome of freedom for Black people. Thus we place the current practice of surveillance of Black people under the same kind of prohibitive reasoning that primitive cultures used to govern against what they viewed as demonic. Antiblackness safeguards taboos that maintain the view of Black people as unclean, foreign and in need of control. A racist mentality operates under the impression that left to their own devices, Black people would cause humanity to regress to basic, uncivilized instincts that destroy the moral and enlightened functioning of society. Freud speculates that originally the punishment for the violation of a taboo was probably left to an inner automatic arrangement, meaning the violated taboo avenged itself. As civilization progressed with further developments of the idea, society took over the punishment of the offender, whose action has endangered his companions. Thus man's first systems of punishment are also connected with taboo (Freud and Strachey 1950); and in multicultural Canada, Black lives are subjected to punitive measures as they are viewed as endangering their companions. Surveillance, management, and racial profiling[2] are then justified as a preventative measure designed to check those who threaten the multicultural ideal through their very nature.

Castration as a form of punishment follows the violation of a taboo. In the construction of culture, which can be defined as the organization of society through prohibition and exclusion from certain spaces, castration results in social and symbolic stratification (James 2010). In spite of the beatifying façade of multiculturalism, we know that little has changed with regards to socially constructed taboos. Castration anxiety is the undeserved psychic suffering of those under the punitive eye of society. The anxiety experienced by Black people within the state is due to a history of trauma and never-ending loss and grief in the quest for freedom. The White gaze threatens the Black body with castration and we contend that surveillance, racial profiling, and management are not tools for justified law enforcement, nor social and economic governance, rather they are prohibitions against perceived taboos. These taboos are based on a cultural order of racism and discrimination that legitimises White supremacy. This statement is also true in Canada despite statements implying slavery was never a part of its history.

According to Laplanche and Pontalis, castration can be identified as a series of traumatic experiences also characterized by an element of loss of or separation from an object (Laplanche and Pontalis 1974, p. 57).[3] Laplanche and Pontalis posit:

> The castration complex has to be understood in terms of the cultural order, where the right to a particular practice is invariably associated with prohibition. The "threat of castration" which sets the seal on prohibition against incest is the embodiment of the Law that founds the human order. (Laplanche and Pontalis 1974, p. 59)

For Black people, freedom remains a prohibited object that is automatically given to all others. The attainment of freedom, as previously mentioned is taboo. Hence, Black lives seeking freedom are threatened and subjected to castration, which in Fred Moten's (2003 p. 177) words is an oft-repeated literal, historical, and material event against people of colour.

To explicate the function of castration in racial discourse, Tapo Chimbganda (forthcoming) highlights the psychic trauma that ensues castration. Leaning on Fanon, she states that the pathogenesis of racialization as a neurosis begins with the rise in tension between what the individual desires as a subject and the accompanying experience of unpleasure that results from a sanctioning of this desire. Fanon (1967) explicates his own analysis:

> The Black man has no ontological resistance in the eyes of the white man. Overnight the Negro has been given two frames of reference within which he has had to place himself. His metaphysics, or, less pretentiously, his customs and the sources on which they are based, were wiped out because they were in conflict with a civilization that he did not know and that imposed itself on him (p. 110).

Ongoing castration anxiety results in a debilitating mental condition that limits the daily efforts of those subject to it. Black lives must contend with the social, psychic, spiritual, and moral repercussions of antiblackness but doing so comes at the cost of their mental health and wellbeing. They live daily with the anxiety of castration because they dare to contend with White supremacy.

The threat of castration maintained through the White gaze can be understood through Fanon's uncanny example when he describes his encounter with a White child who calls attention to him by shouting out "Look, a Negro!" in *Black Skin, White Masks*. Fanon's reaction to the utterances of a small child reveals the castration anxiety Fanon carried with him. The child, or his mother, had no actual authority or power except the White gaze. Surveillance of Blackness is not limited to what would typically be referred to as figures of authority. Rather, surveillance, racial profiling, and management has the potential to be carried out by all who are privileged by White supremacy. Encounters, such as Fanon's reverberate in the Western world where race is founded on the order of White supremacy. In the Canadian context, through mechanisms such as surveillance, Whiteness maintains its strongholds under the guise of lawfulness and policing. State sanctioned surveillance, racial profiling, and management are thus an extension of the White gaze.

Fanon concludes, "I came into the world imbued with the will to find a meaning in things, my spirit filled with the desire to attain to the source of the world, and then I found that I was an object in the midst of other objects ... Sealed into that crushing objecthood." (Fanon 1967, p. 109). Black lives find that they can see only from one point, but in their existence they are looked at from all sides (Lacan and Miller 1991, p. 72). This "being looked at" from all sides results in anxiety, as there is no safe space in which the Black body can be free to exist and function without the prohibition of taboos. Further, Lacan states, "the gaze is presented to us only in the form of a strange contingency, symbolic of what we find on the horizon, as the thrust of our experience, namely, the lack that constitutes castration anxiety" (Lacan and Miller 1991, p. 72–73).

The motivation and intent that informs over-policing and zealous expulsion of Black lives in this contemporary neoliberal moment is the insistence to see "problems" as in need of control and management. According to Stuart Hall "managerialism" is "not only the hallmark of neoliberalism," but "the motor" that drives the set of values, ideas and practices that enables the system to function. "How neoliberalism then gets into the system," he suggests, "is through managerialism" (Hall 2007, p. 111). Surveillance, racial profiling, and management serve as locations of castration; part and parcel of the neoliberal mechanism of Multiculturalism. By framing surveillance, racial profiling, and management of Black lives—consistent practices of Western nations—we make use of the psychoanalytic concept of castration to show the maintenance of social stratification in which Black people are oppressed and White people are privileged.

In summation, we posit that through psychoanalytic language, we can present surveillance as not simply a measure of security, but the unsettlement of Black lives by the White gaze that sees acts of Black empowerment as criminal. By extension, racial profiling is not simply a street level interaction with police; it is the summation of Blackness as a signifier of foreignness and transgression. Management of Black lives is not for the sake of economy, but it is the purposeful and continued control and marginalization of Blackness in the scripted position of endangering alien. These problematic practices are part of a logic and framework that undergirds the manner in which municipalities, cities, and Western nations deal with Black people as 'other' and 'undesirable.'[4] As Baldwin has rightly indicated, when it comes to the practice of surveillance, the only difference between geographical spaces is the tactics used to profile, manage, and ultimately castrate Blackness.

In the rest of the chapter, we examine the surveillance, racial profiling, and management of Blackness beginning with an anecdote recounted by Sam Tecle, who introduces an insidious, all too common Canadian reality for Black people in higher education. Francesca D'Amico examines the management of Black cultural expression and freedom of speech in the media and arts in Canada during the 1990s. Yafet Tewelde explicates on the historical racial profiling and surveillance of Black people during the period of Black activism in Montreal in the 1970s.

"ARE YOU IMPOSTERS?" SURVEILLANCE AND CASTRATION IN THE IVORY TOWER

In what is regarded as one of the most diverse and multicultural universities in Canada, two Black graduate students attend a faculty "Wine and Cheese —Meet and Greet." Though not unusual at this level of academia, these two graduate students are the only Black people in a room filled with professors, department chairs, deans, and other graduate students. Keeping mainly to themselves, the two students seclude to a corner and take time to survey the awkward, but familiar dynamics. Presumably sensing their discomfort, a White woman, also a senior member of faculty approaches them to engage in small talk. The Black students assume it is a gesture to extend collegiality. She comfortably and jovially opens with: "Are you imposters? Are you here for the free food?"

The Black students' experience exposes the ways in which Black students in White ivory tower spaces can be castrated. Castration anxiety caused the two males to stand apart and refrain from mingling even if the purpose of being there was to meet and greet. Within a four-decades long socio-historical context, Multiculturalism often treats the Black Canadian presence with ambivalence. Efforts to affirm Black Canadians as belonging have ranged from reluctant recognition to their erasure from the national memory. Often framed as either 'newcomers' or a 'social problem,' Black

Canadians are often imagined as adjunct to the nation, which then rationalizes the practices of surveillance, racial profiling, and management (Walcott 1997, pp. 118–120, 125). The above anecdote demonstrates that like the nation, the ivory tower is imagined as a 'White' space. As such, within this geography, the Black person is often read as 'new,' 'curious,' 'suspicious,' ultimately as a problem to be dealt with.

In this encounter, the presence of the Black students is marked as perversely different and in violation of a history and legacy of academia as White. These young men appeared to be taking liberties, eating the food, and choosing to remain independent from the rest of the attendees. In this instance, the Black student was cast in the role of transgressor to the traditional (White) social values and norms of academia. Occurrences such as these are not uncommon, as many narratives of the struggles of Black lives within higher education are documented across the Western world, which reinforces our view that castration of Blackness is an oft repeated and present practice by White institutions through the White gaze.

In the example of the two Black graduate students, castration occurs in two ways. First, in the pronounced foreignness of the Black man as threat and violence to the maintenance and preservation of Whiteness. This is often the case in immigrant oppression, which suggests that foreigners are taking away from those who rightly deserve the 'free food.' The general premise that makes the comments of the White female professor permissible is the maintenance of traditional views of citizenships and, by extension, legitimacy in academia as both a traditionally White institution and an annex of the state. The cavalier and normative nature by which this violence is deployed indicates that such comments are made regularly and are in many ways convention. They rely on constructions of taboo and legacies of castration; and they occur so frequently because they regularly go unchecked. These comments indicate that she, the White female professor, is playing on her home turf; and these Black students, simply by virtue of their presence, were foreign and parasitic. Such microaggressions reinforce and reiterate discourses of difference based on belonging and illegitimacy. Within the academy, such encounters are not unique to graduate students. Carl James recounts his own experience as a Black professor and refers to the comments of his students on "how (racialized) ideas influence their expectations of, and interactions with, (him) as a (Black) teacher, and how these in turn have an impact on the teaching and learning process" (James 1994, p. 125). He writes, "Many times, in the course of my work, I have had to pretend I did not hear disrespectful comments, or I would give the impression that I am open-minded enough not to address those comments that challenge my role as a teacher" (James 1994, p. 138). His response, or lack thereof, symbolizes the way comments like these are perceived to be permissible and are constantly repeated. James's example shows the prudence and endurance necessary to guard against the debilitating effects of castration.

Such power dynamics persist with the castration of those perceived as transgressing against tradition. The belief is that the Black person should be present only where she/he serves the objective of delineating White spaces from Black ones. The invisible and hyper-visible paradox contained in the Black body, is referred to by Rinaldo Walcott (1997, 2003) as an absented presence that frames and locks this discourse in constant rotation.

As a Black female professor, Charmaine Nelson also shares her experience of racial violence in academia at the hands of her department Chair, colleagues, and White students. Nelson writes, "it is this mis-recognition the fact that our White students frequently refuse to *see* us as professors and the ways that our bodies (as bodies of authority) become illegible to them, that Black professors must perpetually guard against" (Nelson 2011, p. 118). Collectively, these experiences demonstrate that the so-called colorblind Canadian University, under the guise of multiculturalism, can heighten "the undiminished power of racism and its effect on Black people who continue to comprehend their lives through what it does to them" (Gilroy 1993, p. 102). Events and experiences like these, which are structural and social, historical and cultural, can be read as reminding Black lives that any transgression of the traditional historical scripts that places them as foreign, illegitimate, and endangering of society is taboo. Black students standing apart, eating cheese and wine in a context that signifies white values is a show of freedom that cannot go unchecked. The White female professor's gaze served to express the violation of a taboo. With its legacies of embedded antiblackness and the comforts and protections afforded by the ivory tower, she might as well have shouted "Look two negroes!"

"Can't Repress the Cause (CRTC)": Canadian Hip Hop as Cultural Politics and Blackness Management

Another arena in Canada where Multiculturalism aims to castrate Black peoples is popular culture where music, in particular, has and continues to pose a threat given its inclination towards cultural hybridity (Vannoy

Adams 1996, p. 23). Cultural hybridity challenges the pure Whiteness that surveillance, racial profiling, and management attempt to preserve. The metaphorical miscegenation (Vannoy Adams 1996, p. 23) of cultural hybridity in music was rendered problematic, because it fundamentally threatened Canadian citizenry. As a consequence of its attempt to challenge and revise state supported notions of Canadian identity, an identity that had long been synonymous with Whiteness,[5] Canadian Hip Hop was managed through antiblackness. In order to demonstrate the antiblack elements of the Canadian music industry, an examination of state and music management practices reveal an inclination among those in power to control Canadian popular culture by reinforcing and reifying White supremacy through the threat of castration and the White gaze.

Popular culture in Canada has been managed through the setting up of governing bodies specifically geared towards Blackness management, the first of which is the Canadian Radio-Television and Telecommunications Commission (CRTC).[6] The apathy of radio station programmers towards domestic music led to a campaign intent on encouraging the development, promotion, and exposure of Canadian content and talent. Following a lengthy public hearing process, the CRTC initiated an elaborate series of broadcasting policies in 1971, including a 30% radio quota intended to maintain Canadian cultural "integrity" amid American popular culture hegemony (Edwardson 2009, pp. 139–142, 147, 155; Warner 2006, p . 49). These decisions particularly highlighted an attempt to deflect the critique that the marketplace reflected a lack of access and diversity of offerings, and isolated Canadian artists, particularly racialized ones, to isolated genre markets/format radio (Edwardson 2009, p. 149; Warner 2006, p. 49). Part of the logic informing isolated markets was the notion that Black popular culture and artistry was taboo and castrating measures were needed to preserve *Canadianism*, which is in effect, a reification of Canada as a White space.

Black Canadian artists resisted this control through acts of artistic protest and business strategy, acts interpreted as radical and therefore violating fidelity and submission. As a result, these artistic actions led to a series of new policies and laws that came to be used to justify the silencing of these Black artists. Though multiculturalism was supposedly intended to usher in a new social contract, Hip Hop's antagonistic posturing did not fit comfortably into that fantasy in that if Black music became Canadian within the multicultural paradigm, the sovereign would become alienated from Whiteness. Simply put, this would be the violation of a taboo, an endangering of Whiteness, which could not be allowed to happen. With the rise of Hip Hop in this new multicultural Canadian paradigm, divisibility became a real threat to notions of sameness, which necessitated the application of policies and laws to undermine any real change. As such, Canadian broadcasting, an arm of Canadian Whiteness and the White gaze, created a subdivided and specialty niche-market that led to the problematic collapse of all 'Black music' into one genre.[7]

Anthropologist Remi Warner contends that Canadian format radio, defined largely by perceived audiences in segmented markets, unwittingly invented "narrowcasting." This marginalizing technique conveniently satisfied political concerns of employing Canadian popular culture as a nation-building tool intent on achieving diversity of choice and access. However, it also castrated Black popular culture presence by highlighting the perceived invisibility of Black Canadians even while Multiculturalism conceives of these subjects as the nation's (un)official public "core culture." Consequently, narrowcasting was, and continues to be, framed through unequal power relations, structural inequities, and managed visibility characteristic of a hegemonic, liberal-multiculturalist, late capitalist state formation (Warner 2006, pp. 49, 59–60).

The castration of the Black presence within Canadian popular culture highlights the paradoxical and problematic nexus between identifying and servicing specialty markets and audiences, while attempting to incorporate bodies identified as 'other.' Practitioners point to the adoption of the term 'urban music' as evidence. They claim the term attempts to represent an array of music created and performed by Black artists for Black people. In reinforcing this racial classification, the term contains and marginalizes Black Canadian contributions to a nondescript music-ghetto. Practitioners contend that the term permits the surveillance, racial profiling, and management of Black lives and their artistic contributions, while simultaneously engendering their hyper-invisibility by ambiguously ghettoizing all forms of Black artistry into sameness while simultaneously separating Black sameness from Canadian sameness.

Cultural Critic Dalton Higgins argues that, "the 'urban' designation (is intended) to describe away rap music, RnB, Soul…(it) is a more polite way of saying Black." Higgins contends that the term and radio format, first introduced by African American radio broadcaster Frankie Crocker, was employed as a corporate strategy to elicit advertising revenues and partnerships. Like DJ Mel Boogie and Tony 'Master T' Young, Higgins argues that the term is problematic for its ambiguousness and because it was not created by practitioners. Rather, industry gatekeepers created the term in order to make Black contributions more palatable (Performing Diaspora 2013).

In order to meet the rules that govern Canada's state project of inclusiveness, the state's management of Blackness reflects a paradoxical practice which is undergirded by the desire to eliminate Blackness while also incorporating and commodifying it.[8] It is in this liminal space where Blackness becomes, under management and surveillance, palatable. Take the examples of radio and Juno recognition,[9] in 1990, Milestone Radio first petitioned the CRTC for an 'urban' format station (Nazareth and D'Amico 2012). The decision, a controversial one that ended in Milestone losing their bid to Country music radio, was greeted with antagonism in the form of a 1991 protest record titled "Can't Repress the Cause (CRTC)" (a take on the CRTC acronym). The recording, which featured a number of popular Black Canadian artists, including reggae artist Messenjah, R&B singer Devon, and rappers Maestro Fresh Wes and Michie Mee, cited the contradictory phenomenon of receiving industry accolades despite (none or very little) radio rotation support. In the recording Maestro rapped, "it's a trend that keeps us trapped, we pumpin' out LPs of gold, still local radio ain't got no soul" (Can't Repress the Cause 1991). The second example, Juno recognition for Black artistry gained scrutiny in 1998 when Rap trio Rascalz refused to accept the Juno for best rap recording at the 27th Annual Awards following the organizers' decision not to include the Rap, Reggae, and Dance awards in the telecast (LeBlanc 1998). In a prepared statement, the Rascalz stated, "In view of the lack of real inclusion of Black music in this ceremony, this feels like a token gesture towards honoring the real impact of urban music in Canada." The group's co-manager Sol Guy added, "Urban music, Reggae, R&B, and Rap, that's all Black music, and it's not represented (at the Junos). We decided that until it is, we were going to take a stance" (LeBlanc 1998).

This practice of Blackness management in popular culture has had a number of debilitating consequences. According to Brathwaite and Branker of *The Northside Research Project*,[10] across Canada's five major music markets (Edmonton, Halifax, Montreal, Toronto, and Vancouver) artists cited the lack of a nationally recognized infrastructure as the major source of frustration. For Higgins, this lack of industry support in terms of radio and record label promotion has been a historical reality for Black artists, particularly those who identify as Hip Hop (Higgins 2012). Rapper Wes 'Maestro' Williams confirmed this when he claimed that, "there (is) not really a (Hip Hop) industry here (in Canada), there never really was." Moreover, even in instances of international success, as in the case of Drake and K'Naan, Maestro maintains that, "those (artists) are all anomalies, (and) that's not the norm." (Wes "Maestro" Williams 2012). According to *Northside* participants, the lack of support and infrastructure has come to be defined as: the inability to receive adequate video and radio exposure, geographical inaccessibility across regionalized markets, an inability to access venues as a consequence of stereotyping, a lack of funding support/inaccessible knowledge regarding funding, a deficit of effective management and business training to create a genre-specific strata of industry professionals, and a debilitating fragmentation to sustain a Hip Hop scene (Brathwaite and Branker 2006, pp. 5, 22).

The *Northside Research Project* participants argued that these inequities are grounded in racist perceptions and practices. Practitioners contended that the Canadian mainstream media reified a myopic portrayal of Hip Hop that pigeonholed the genre as an inherently criminal practice, rather than an art form, rooted in poor Black communities and supported by the disenfranchised Black urban media (Brathwaite and Branker 2006, pp. 21–23, 26). These perceptions, rooted in racist notions of identity, resulted in a series of racial, cultural, and economic disparities and barriers to mainstream exposure (Brathwaite and Branker 2006, pp. 5–6, 21). For these artists, Black musical expression was a quest for freedom, which would endanger the Canadian identity constructed through White ideals. Participants argued that the institutional apparatus for the arts in Canada lends far greater support to alternative rock culture (a genre of music mostly perceived as White), and that this was perhaps most evident in national funding models. Participants argued that non-White grant applicants with less knowledge of the funding process are immediately at a disadvantage, because they lacked the structural support by way of mentorship in career management. By limiting resources and funding for Black art, the White supremacist status quo is maintained. Moreover, participants noted that in order to successfully acquire grant support, or increase their chances, they needed to modify the description of their work (for example, the decision not to use the term 'Hip Hop'), which ultimately, is another layer of castration (Brathwaite and Branker 2006, pp. 21, 25).

"I Found Something in Canada that I Had not Found Before": Multiculturalism and the State Surveillance of Blackness

In the same year Canada implemented its Multiculturalism Policy (1971), the Canadian government also intensified its monitoring and infiltration of Black Canadians, specifically social justice groups who outlined an explicit Black Power agenda to engage in a myriad of initiatives in Toronto and Montreal to challenge Canada's anti-Black racism (Stasiulis 1989). The Royal Canadian Mounted Police (RCMP) responded by partnering with the US Federal Bureau of Investigation (FBI) who loaned their Black agent, Warren Hart, to dismantle Black organizing and other undesirables through a pseudo spy exchange program (Hewitt 2002). Hart operated within the FBI's notorious Counter Intelligence Program (COINTELPRO), (Churchill and Vander Wall 1990) which was theoretically decommissioned in 1971, yet appeared to remain active enough to influence the RCMP and Canada in quelling its own "Negro problem."[11]

Problematized by Canadian authorities, Black empowerment was used as a justification for surveillance, which placed social justice activists under the intensity of the White gaze. The criminalization of those fighting for the citizenship of Black lives was based on traditional scripts of Blackness as "radical" and therefore transgressing against Canadian ideals based largely on White citizenship. Consequently, Multiculturalism juxtaposed with the infiltration of Black social justice movements underpinned a castration campaign against Black lives to protect Canada's White supremacy order. When justified through calls for nationalism and security, Canada's inclusivity and acceptance discourse therefore acts as a code for eliminating challenges to White hegemony. Black lives, and by extension Blackness, as an explicit challenge to White supremacy, must pledge fidelity to a State that recognizes them not as citizens, but as subjects necessitating the elimination of their attempts for justice.

The protest at Sir George Williams University (now Concordia University) in Montreal in 1969 illustrated the height of the militancy of Black Canadians.[12] The RCMP had a very deep interest in monitoring the famous "Computer Centre occupation" occurring in Montreal at the time. They decided that the poor information they had was due to a combination of factors: the restrictions on carrying out surveillance and investigative activities on campuses, specifically the recruitment of sources, and a fear of having these activities at universities exposed in the media and; the predominance of Whites in the force, which made covertly monitoring Black student groups difficult. It was this latter factor, in particular, that led the RCMP to request the services of agent provocateur Warren Hart (Hewitt 2002, p. 151).

While Hart's activities also included spying on anyone deemed beyond the norm of political engagement, within Black political organizations, Hart's role was also to discredit the activities of Black activists by pressing them to engage in violent activities against the State.[13] Although many refused Hart's proposals to drive in armed convoys or bomb government buildings (McQuaig 1981), Hart was able to take advantage of Black youth desperate to fight racist White Canada. For example, Richard Atkinson, famed leader of the "Dirty Tricks Gang" bank-robbing group from the 1960 and 1970s (Tripp 2011) recounts:

> My political ideology at the time of being wooed into Black Panthers in Toronto via Warren Hart was to bring to the world a conscious reality of the racial inequality that existed in North America. And I was ready and willing by any means necessary to create change in the status quo. If that meant blowing up a building, a police car, or hijacking a plane to bring to light the reality of racial inequality, so be it…I was convicted of robbing a bakery to feed the Black Panther defense fund and I used Warren Hart's gun. This is the very same weapon that the Canadian government allowed him to have in his role as agent provocateur (Atkinson, "Ricky's Blog" [online]).

These radical young people were rallying against being contained in identities of foreignness and transgression. These are the very weapons by which the authorities chose to entrap them. With Hart urging them to transgress, Blackness was seen as fulfilling anti-Canadian tropes justifying the White gaze, and therefore castration. Quests for social justices were curtailed by the enforcement of punitive justice.

Unfortunately for the RCMP, even their recruited Black agent fell under the threat of castration through the white gaze. Hart's activities under the RCMP became so unruly that the Canadian government was forced to launch the *Commission of Inquiry Concerning Certain Activities of the Royal Canadian Mounted Police* in 1977. Hart's role became fodder for the mainstream news. This in turn led to an interview in 1978 with the Canadian English language broadcast station CTV Television Network that exposed his role within the RCMP. Hart felt betrayed by his employers (the Canadian government) who forced him to take the position of key witness at the inquiry and then ironically refused him permanent residency as he had requested when his work with them was complete. Ultimately, his job as a Black man was to maintain the dominance of Whiteness, and his request for Canadian status was counterproductive to this endeavour. Hart was deported and allowed to return periodically only to testify on his involvement in the sabotaging of Black activism, in effect highlighting the castrating otherness of Black people within the Canadian landscape. The inquiry did not call on any of the Black political organizers who Hart mentions as his key targets, which allowed him to testify unchallenged. This was especially troubling because one of Hart's primary objectives, according to those he monitored, was to encourage and incite violent actions by, and amongst, Black and Indigenous political organizers.[14] In effect, his purpose was to monitor, discredit, and ultimately destroy Canada's Black Power Movement through the act of surveillance. Hart's spying and infiltration was used to maintain the script through which White supremacy seeks to deny Black agency. Through these acts of surveillance, these groups were presented to Canada as endangering the State. The detailed files collected by police organizations allowed the State to use their already preconceived political conclusions as legitimate judicial premises for the indictment of radical political prisoners (Boyce Davies 2008, p. 194).

Within the logic of the Canadian state, the policing of Black people engaged in political organizing is a method of castrating "deviant" and "threatening" Black bodies in order to maintain and preserve Whiteness. This act highlights the difficulty Canada has in according Black people socio-political and economic equality within the current logic structure of White supremacy. The project of Multiculturalism uses the language of diversity and tolerance to maintain Canada's current power structure, while rebranding it as part and parcel of Canadian cohesion and national unity.[15] Within this discourse, Black bodies pose a threat for whiteness and this fact necessitates that only tokenistic policies, such as those contained in Multicultural ideology and policy, can be tolerated (Welsing 1991, p. 9). This history also illuminates how the current police practice of racial profiling must be contextualized within the longer history of spying on Black political organizers and communities. Both are part of the larger objective of targeting, monitoring, and disciplining those who challenge the White hegemonic power structure. Canada's reaction to Black movements of the 1960 and 1970s makes clear that this form of Blackness cannot be subsumed into the existing logic of tokenism within Canadian society.

CONCLUSION

The acts and threats of castration on Black people is the curtailing of social justice in Canada. What happened in the 1970s with the introduction of its Multiculturalism policy and the intentional stifling of the rise of Black social justice groups has had long standing effects. Since its inception, Multiculturalism has proven to disproportionately benefit White people, though different interest groups at diverse moments in Canadian history have benefited by virtue of becoming White. White privilege is a process of inclusion over time, as many historians and sociologists have argued, people 'become' White. Not all Whites have been equally positioned or privileged and while it is not true that Multiculturalism in Canada has *only* benefited Anglo-descended White people, the great and disparaging disproportion of those bodies that have become instantiated *as* Canadian are overwhelmingly White rather than Black. These are the same people that are privileged in the arts and in generally every other sphere of freedom and liberty within the nation. Rousseau in *The Social Contract* made the observation that impunity is reserved for citizens, who by this virtue cannot be considered to transgress against the Sovereign, regardless of their actions. Citizens, as legitimate to the State, cannot have any interests contrary to the State (Rousseau 2006), therefore do not operate under taboo laws. Any violation to the law of the Sovereign is deemed to come from outsiders, which is why Black lives become contained in discourses of foreignness and illegitimacy, effectively alienating them from the State. By force, the State creates rights and laws, which are maintained by castrating mechanisms, ensuring the power and authority granted only to citizens remains with those who fit the description.[16] These rights and laws ensure that citizens can transgress legitimately. This, in turn, results in being justified or legalized. In contrast the subject transgressions are considered taboos and that is how the quest for freedom and social justice become criminalized.

Between the United States and Canada scholars have made multiple distinctions in the way race has emerged historically and socially. These distinctions have supported fantasies of Canada as an oasis of racial equality and Multiculturalism, a country where diversity through unity and purposed sameness affords all cultures flourishment and success. However, according to Walcott, Canadian fictions of sameness have concealed and denied otherness within the nation, while rendering acceptable the continued disenfranchisement of racialized Canadians. Bannerji concurs with this statement, contending that racism has become so naturalized and pervasive that it has become invisible to those who are not adversely impacted by state-sanctioned practices.[17] Critics contend that re-imagining Canada requires the presence of "others" who create counter-hegemonic discourses to expose Canada's hidden and secretive histories, social relations and the racialized code that produces Canada as a 'White nation.'[18]

Through the implementation of multiculturalism, the Black person experiences castration anxiety, because the threat of castration is always imminent and a looming presence that curtails the daily freedoms in the lives of Black people. Freedoms others take for granted. As (Baldwin et al. 1989, p. 45) states, castration occurs in many cases where threats are followed through with actions, such as deportation, incarceration, and police brutality. Other forms of penance are prescribed onto the Black body, which we witness in the news and in our media where Black lives are used to signify transgression. Multiculturalism as a tool, therefore maintains White supremacy as long as Blackness signifies otherness. Black lives are considered subjects, not citizens in Canada. Therefore, they remain under the forceful prohibition of the State. The power imbalance is maintained through surveillance, racial profiling, and management with the White gaze continually issuing the threat of castration. Due to the forced illegitimacy of Black lives in Canada and under the guise of Multiculturalism, selection between subject and citizen has been maintained.

For Bannerji, Canadian Multiculturalism has always been a site for struggle and contestation in which the state's formation depends upon the conquering imagination of White supremacy, and a legacy of survival anxieties and aggressions in the form of colonialism, conquest, and exclusionary tactics.[19] Therefore, Multiculturalism, through castration, is a form of systemic exclusion of Black people and the maintenance of difference between races through castration. The taboo prohibited by Multiculturalism is the ascendance of Black people into true citizenship and into the acceptance of things deemed "Black" as Canadian. Consequently, in the social strata that we know as the Canadian multicultural narrative, the true citizens are White, and they reign supreme in all things Canadian. While Canadian immigration policy has long indicated approaches to belonging and nationhood, Multicultural policy framed and codified in law who and how one officially belonged to the nation. The policy endeavours to subordinate and produce inside/outside binaries by textually etching in non-Whites as not Canadian, while simultaneously carving them out as tangential to the nation (Walcott 2003). Consequently, this policy reinforces the notion that Black people will never—in actuality if current logic holds—can never meet the standard of Canadian citizenship. As we have observed, the quest for the same rights and freedom is viewed as tangential to Multiculturalism, placing Black lives in a persistently liminal and paradoxical situation. Being born in Canada, living and dying in Canada, and wanting to be Canadian does not a Canadian make!

NOTES

1. For further reading see: Mackey, *The House of Difference: Cultural Politics and National Identity in Canada* (New York: Routledge, 1999); Yasmeen Abu-Laban and Christina Gabriel, *Selling Diversity: Immigration, Multiculturalism, Employment Equity, and Globalization* (Toronto: Broadview Press, 2002); Himani Bannerji, *The Dark Side of the Nation: Essays on Multiculturalism, Nationalism, and Gender* (Toronto: Canadian Scholars Press, 2000); Nira Yuval-Davis, "Some Reflections on the Question of Citizenship and Anti-Racism." In *Rethinking Anti-Racisms: From Theory to Practice* edited by Floya Anthias and Cathie Lloyd (New York: Routledge, 2002, pp. 44–59) (Abu-Laban and Gabriel 2002; Bannerji 2000; Mackey 1999; Yuval-Davis 2002).
2. The Ontario Human Rights Code definition of racial profiling, "is any action undertaken for reasons of safety, security or public protection, that relies on stereotypes about race, colour, ethnicity, ancestry, religion, or place of origin, or a combination of these, rather than on a reasonable suspicion, to single out an individual for greater scrutiny or different treatment." http://www.ohrc.on.ca/en/what-racial-profiling-fact-sheet.
3. The object normally symbolizes the desired phallus, which is not an actual biological unit, but a manifestation of power and authority. The phallus is non-gendered in contemporary psychoanalytic theory and though Freud spoke in what are perceived as gendered terms, it is important to remember that the unconscious is not gendered, not racialized and not cultured until it begins to function under the threat of castration within the social order.
4. The recent lack of focus and attention toward the plight of Black African refugees is very insightful of this point.
5. We assert that Canadian identity is still presently very much synonymous with Whiteness.
6. Unlike the Federal Communications Commission (FCC), its closest American equivalent, the CRTC is not an independent agency of the federal government. Rather it is a public organization that regulates national broadcasting and telecommunications only. Whereas the FCC oversees the media, access to broadband, public safety and homeland security.

7. This genre has retained many designations over time and is currently referred to as 'urban' music.

8. In further highlighting the congruency between the Canada and the US, see Hooks, B., & West, C. (1991). *Breaking bread: Insurgent Black intellectual life.* Toronto: Between the Lines where the authors make the point that Blackness is managed through the tension of being simultaneously despised (fear of inclusion) and desired (intensely commodified) in the American popular culture context.

9. The Juno Awards are presented annually to Canadian musicians in order to acknowledge their artistic and technical achievements. Winners are chosen either by members of the Canadian Academy of Recording Arts and Sciences or a panel of experts (this largely depends on the nature of the award). In almost all of the major categories, nominees are determined by market sales during the qualifying period, and in genre-specific categories, nominees are determined by a panel of experts.

10. In 2006, Canadian poet/rapper Wendy "Motion" Brathwaite and journalist Saada Branker were approached by then-Equity Coordinator at the Canada Council for Arts Anthony "Nth Dgri" Bansfield, to explore the nature of Hip Hop recognition and support throughout Canada, both historically and contemporaneously, through sample focus groups.

11. Austin, *Fear of a Black Nation*, shows this was the language used by the RCMP when discussing Black people.

12. On January 29, 1969 over 400 students occupied the university's computer lab to protest the university's mishandling of racism allegations against a professor at the school. The protest lasted until February 11 when negotiations failed to bring about an end to the stand-off and riot police stormed the lab. A fire broke out in the lab causing an estimated two million dollars in damages and 97 being arrested.

13. Hart Interview with Jim Reed *Prime Time*; Austin, *Fear of a Black Nation*.

14. McQuaig, "The Man with the Guns".

15. Mackey, *The House of Difference*; Abu-Laban & Gabriel, *Selling Diversity*; Yuval-Davis, "Some Reflections on the Question of Citizenship and Anti-Racism"; Bannerji, *The Dark Side of the Nation*.

16. The term "fit the description" is a phrase synonymous, almost mythologized, with encounters of racial profiling by the police.

17. Bannerji, 114.

18. Walcott, 116; Bannerji, 120.

19. Bannerji, 92–97, 105–107.

References

Abu-Laban, Y., & Gabriel, C. (2002). *Selling diversity: Immigration, multiculturalism, employment equity, and globalization.* Toronto: Broadview Press.

Ahmed, S. (2004). Affective economies. *Social Text, 22*(2), 117–139.

Austin, D. (2013). *Fear of a black nation: Race, sex, and security in sixties montreal.* Toronto: Between the Lines.

Baldwin, J., Standley, F. L., & Pratt, L. H. (1989). *Conversations with James Baldwin.* Jackson: University Press of Mississippi.

Bannerji, H. (2000). *The dark side of the nation: Essays on multiculturalism, nationalism, and gender.* Toronto: Canadian Scholars Press.

Brathwaite, W., & Branker, S. (2006). *The northside research project: Profiling hip hop artistry in Canada.* Presented to the Canada Council for the Arts, Ottawa, ON.

Browne, S. (2015) *Dark matters: On the surveillance of blackness.* Durham and London: Duke University Press.

Boyce Davies, C. (2008). *Left of Karl Marx: The political life of black communist Claudia Jones.* Durham: Duke University Press.

Chimbganda, T. (Forthcoming). *Privileged space: A psychoanalytic paradigm for social justice in education.* Maryland: Rowman and Littlefield.

Churchill, W., & Vander Wall, J. (1990). *The COINTELPRO papers: Documents from the FBI's secret war against domestic dissent.* Boston: South End Press.

Dance Appeal—Can't Repress the Cause (CRTC). (1991). *Can't repress the cause (CRTC).* Single (FLIP—SOM 1235, 1991).

Du Bois, W. E. B. (1904). *The souls of black folk: Essays and sketches.* Chicago: A. C. McClurg.

Edwardson, R. (2009). *Canuck rock: A history of Canadian popular music.* Toronto, ON: University of Toronto Press.

Fanon, F. (1967). *Black skin, white masks.* New York: Grove Press.

Freud, S., & Strachey, J. (1950). *Totem and Taboo: Some points of agreement between the mental lives of savages and neurotics.* New York: WW Norton & Company.

Freud, S., & Strachey, J. (2000). *Three essays on the theory of sexuality.* New York: Basic Books.

Gilroy, P. (1993). *Small acts: Thoughts on the politics of black cultures.* London: Serpent's Tail.

Hart, W. (1978). *Prime time.* Hosted by Jim Reed. Toronto: CTV, Television.

Hewitt, S. (2002). *Spying 101: The RCMP's secret activities at Canadian universities, 1917–1997.* Toronto: University of Toronto Press.

Higgins. (2012, December 8) *Oral interview conducted by Francesca D'Amico for "An evening with Dalton Higgins,"* Toronto, Ontario.

James, C. E. (1994). In James & Shadd (Eds.), *I've never had a black teacher before in talking about difference: Encounters in culture, language and identity* (pp. 125–141). Toronto: University of Toronto Press.

James, C. E. (2010). *Seeing ourselves: Exploring race, ethnicity and culture.* Toronto: Thompson Educational Press.

Lacan, J., & Miller, J. (1991). *The seminar of Jacques Lacan* (Norton paperback ed.). New York: W. W. Norton.

Laplanche, J., & Pontalis, J. B. (1974). *The language of psycho-analysis.* New York: Norton.

LeBlanc, L. (1998). Rascalz refuse award to protest Junos. *Billboard, 110*(4), 54.

Mackey, E. (1999). *The house of difference: Cultural politics and national identity in canada.* New York: Routledge.

Moten, F. (2003). *In the break: The aesthetics of the black radical tradition.* Minnesota: University of Minnesota Press.

McQuaig, L. (1981, June 13). The man with the guns. *Ottawa Citizen: Today Magazine.*

Nazareth, E., & D'Amico, F. (2012). Urban music. In *The Canadian encyclopaedia.* Retrieved May 2, 2012 from http://www.thecanadianencyclopedia.com/en/article/urban-music-emc/.

Nelson, C. (2011). Toppling the 'great white north': Tales of a black female professor in canadian academia. In S. Jackson & R. G. Johnson III (Eds.), *The black professorate: Negotiating a sabitable space* (pp. 108–34), New York: Peter Lang.

Roundtable 2: Toronto Music Industry Insiders. *Performing diaspora 2013: The history of urban music in Toronto.* Retrieved June 1, 2013 from http://tubman. info.yorku.ca/activities/performing-diaspora/performing-diaspora-2013/performing-diaspora-2013-videos/.

Rousseau, J. (2006). *The social contract.* New York: Penguin Books.

Stasiulis, D. K. (1989). Minority resistance in the local state: Toronto in the 1970 and 1980s. *Ethnic and Racial Studies, 12*(1), 63–83.

Tripp, R. (2011, April 9). Feature: Last chance for aged thief. *The National Post.*

Vannoy Adams, M. (1996). *The multicultural imagination: "Race," color, and the unconscious.* London: Routledge.

Walcott, R. (1997, 2003). *Black like who?: Writing Black Canada.* Toronto: Insomniac Press.

Warner, R. (2006). Hiphop with a Northern Touch!? Diasporic wanderings/wonderings on Canadian blackness. *Topia, 15,* 49 (Spring).

Welsing, F. (1991). *The Isis papers: The keys to the colors.* Washington: CW Publishing.

Wes "Maestro" Williams, *Oral interview conducted by Francesca D'Amico, on February 1 2012, Toronto, Ontario.*

Yuval-Davis, N. (2002). Some reflections on the question of citizenship and anti-racism. In F. Anthias & C. Lloyd (Eds.), *Rethinking anti-racisms: From theory to practice* (pp. 44–59). London: Routledge.

AUTHOR BIOGRAPHIES

Sam Tecle is a Ph.D. student in the Sociology Department at York University in Toronto, Canada.

Tapo Chimbganda recently completed her Ph.D. in Education at York University in Toronto, Canada.

Francesca D'Amico is a Ph.D. student in the History Department at York University in Toronto, Canada.

Yafet Tewolde is a Ph.D. student in the Social, Political Thought Department at York University in Toronto, Canada..

CHAPTER 12

Sousveillance as a Tool in US Civic Polity

Mary Ryan

HISTORY OF SOUSVEILLANCE

The monitoring of human behavior, or activities, for the purpose of influencing, directing, or protecting people has been practiced in human cultures for centuries. The formal study of surveillance itself dates back to Jeremy Bentham's 1791 publication of the seminal *Panopticon*, which foresees architecture based on surveillance in places like schools, prisons, and workplaces (Bentham and Bozovic 1995). Bentham suggested that the use of total surveillance would prevent people from misbehaving and increase self-regulation. By "total," he included three kinds of surveillance: actual, implied, and potential. As this collection has shown, Foucault (1977) describes the Panopticon as a method of social control where people regulate their own behaviour by conforming to social rules exerted by those in power. Surveillance is not always a top-down practice by government officials, however; sometimes the process of self-regulation can occur more from the bottom up and hold government officials, or normative powers, accountable.

According to Vian Bakir's research, sousveillance was developed from Steve Mann's analysis of Harold Garfinkel's 1967 ethno-methodological

M. Ryan (✉)
Department of Political Science, Virginia Tech, Blacksburg, USA
e-mail: maryryan@vt.edu

© The Author(s) 2017
S. Flynn and A. Mackay (eds.), *Spaces of Surveillance*,
DOI 10.1007/978-3-319-49085-4_12

approach to breaching norms (Bakir 2010, p. 16). Sousveillance emerged from research by Mann in the 1980s before hand-held computers, digital cameras, and smart phones were commonplace. By developing technologies like WearCam and WearComp, Mann directly connects sousveillance technology with personal agency and power (Mann et al. 2003). He promotes personal empowerment through the restoration of individuals over the technology they use (Mann 2001). The devices allow a user to watch, record, and broadcast her surroundings. Importantly, sousveillance enables people, in Mann's vision, to stand against the state's operations of surveillance.

The legal and formal structural challenges to sousveillance in American politics in the twenty-first century make sousveillance a timely and important topic in civic life. The U.S.'s surveillance shift is largely attributable to the War on Terror and underscores how political and strategic communications centralized the necessity of surveillance in order to alleviate the fear of terrorism (Bakir 210, p. 20). Most famously, in 2001, the U.S. Congress passed the USA PATRIOT Act, which is an abbreviation for Uniting and Strengthening America by Providing Appropriate Tools Required to Intercept and Obstruct Terrorism. This law permitted increased government surveillance of financial and personal records, phone wiretaps, and email and internet surveillance. Sousveillance in direct conversation with this idea of a surveillance society. Mann describes the role of sousveillance as a kind of reflection, meaning the philosophy and procedures of using technology to mimic and challenge bureaucratic organizations (Mann 2001). Sousveillance turns surveillance on its head: individuals now observe the organizational or bureaucratic observer. Sousveillance redirects the mechanisms and technologies previously held under exclusive control of the government. Ultimately, Mann's use of sousveillance increases the equality of opportunity to practice surveillance, thus bridging the gap between the surveiller and the surveillee, allowing for greater equality, civic freedom, and personal empowerment. As Mann (2004) describes:

> The term 'sousveillance' refers both to hierarchical sousveillance, e.g. citizens photographing police, shoppers photographing shopkeepers, and taxi-cab passengers photographing cab drivers, as well as to personal sousveillance (bringing cameras from the lamp posts and ceilings, down to eye-level, for human-centered recording of personal experience) (620).

In this description, Mann seems to suggest that sousveillance not only has social value, but that it is a more humanistic approach to surveillance. The potential for sousveillance to better the value of surveillance in democratic society is bolstered by the personal, humane application of technology. There is no guarantee this positive potential will be realized, which is why technology and surveillance must be construed as sociopolitical tools to be manipulated by people and not mechanical tools which rule them. This chapter will turn now to a fuller examination of the many possibilities and cautions inherent in sousveillant technologies which have already been documented in various artistic projects.

Art and Sousveillance

Art can help society imagine what the future could look like. Creativity contributes to an "informed knowledge society" which is critical to cultivating a socially responsible society (Mann et al. 2003, p. 333). As the United States grows and transforms, artistic vision warrants a place in democratic thought. Wearable art, in particular, plays a significant and inventive role in sousveillance culture, even if it only impacts overt legal and political struggles in indirect ways. For the purposes of this chapter, wearable art is a technological invention, which is worn on or adhered to one's body and creates a product which can also be displayed artistically. Many of these inventions performed a singular, or limited, number of functions at the time of their creation; today, there are technologies on the market, such as the Narrative Clip device created by the Swedish company Narrative in 2012, which unify several inventions as well as offer substantial memory storage and sharing capacities.

The conceptual media artist Alberto Frigo is a prolific wearable artist, who is also, like Mann, a developer of sousveillant devices (Sterling 2006). However, Frigo's intentions are distinctly different from Mann's. Where Mann is interested in power dynamics within society, Frigo was interested in understanding himself better. Consider Frigo's most intense initiative. Frigo is currently in the midst of a project that began in 2004 and is slated to end in 2040, which is geared toward tracking his own actions, something which he has dubbed lifelogging (Wilson 2015). In this process Frigo tracks everything that his dominant hand, the right hand, has used. This introspective form of documentation has helped him to not only understand himself better but to more effectively connect with the world around him. As the societal dynamics for healthy democracy are constantly cycling through discussion, Frigo's sousveillant technology should not be dismissed as selfish or insufficient. Indeed, in an individualistic society like the United States, technology, which can improve the depth and authenticity of our communication and relationships with other people, may prove quite fruitful. Frigo's invention demonstrates how art and sousveillance can go hand in hand to document and explain people's individual experiences. This can be seen as a kind of personal history which is tantamount to telling a social truth of one's own lived experience.

Mann has two other key inventions from 2001 which function as a kind of protest art and social commentary. The singular function of these devices illuminates the reasons sousveillant technology was initially created; although society now has access to devices capable of multi-tasking and engaging in technical feats not rooted in sensitivity to power imbalances, these devices remind users that the foundation began with a different intent. Initially, the wearable interactive art piece called the HeartCam was invented with this in mind (Mann 2001). This device reverses the male gaze, suggesting that some technological inventions are rooted in sexual harassment experienced by users in daily life. Companies like Nestle Fitness have further developed this concept into a bra which used the vantage point of the breasts as the camera. Nestle utilized the "Bra Cam" during Breast Cancer Awareness Month to encourage women to conduct self-examinations of their breast health (Stampler 2014). By this measure, sousveillant technology moves from an individual's usage to a communal reminder, demonstrating one way that sousveillant technology can bring people together for social change and be used as a tactic in larger social or political campaigns. Next, he created what he calls an invisibility, or aposematic suit, which is directly correlated to self-defense. The suit has two modes of operation: It can be used like a chameleon's cloak to appear invisible and hide from prey; during this mode where the suit becomes transparent, the video display shows what is behind the person wearing it. This gives the viewer, or other people in close proximity who are not wearing the suit, the illusion that they can see through the individual wearing the suit. In the other mode, the suit becomes reflective like a mirror, hence the aposematic name, to deter prey. Here, the person wearing the suit resembles a two-sided video mirror. Thus, a potential attacker sees himself displayed on the suit itself. This reflection occurs whether the attacker approaches from behind or from in front of the individual wearing the suit. This process is comparable to department store video displays of surveillance camera feeds at the front of the store, or in checkout lines which alert potential shoplifters to the fact that they are being watched. Yet, sousveillant technology takes the traditional surveillance video a step farther by placing an individual in control of how they will be seen, if at all. The inventor Shinseungback Kimyonghun has further enhanced the aposematic jacket by adding a vocal feature to the wearable computer camera (Mann et al. 2003). The lenses on Kimyonghun's jacket emit an audible warning which states "I can record you" to thwart a possible attack. Should the person wearing the suit push a button, the jacket records her surrounding scene in panoramic views and transmit the images to Web. This ability to incite community awareness, help other members of the community, and endure challenges in a public setting, not in private isolation, are important for a democratic society.

Numerous fiction books also contribute to visions of sousveillance cultures, primarily through the genre of science fiction books. David Brin's 1990 novel *Earth* shows citizens equipped with augmented reality gear called Tru-Vu goggles, as well as cameras which display reciprocal accountability between citizens and the book's primary authority figures, law enforcement officers. Robert Sawyer's 2002–2003 trilogy, *Neanderthal Parallax* shows the species the Homo neanderthalensis living in a parallel universe. Each individual has a recording and transmission device mounted on her forearm. Their lives are continuously monitored and stored in an alibi archive. This archive is only accessible by the owner, or pertinent authorities, in the event of a criminal investigation. Similar to the *Parallax*, in John Crowley's (1985) short story "Snow", individual lives are also recorded, but in this story, the recordings or memories, are available to any interested party —presuming they can afford to purchase them. In "Snow", there is a suspended camera which records a person's entire life and then sells it as a consumer product so families can document the memories for future generations. As one man combs through 8000 h of video from his now deceased wife's life, he finds the technology does not sidestep his personal grief. Crowley and Sawyer's stories reveal that technology cannot ascribe value for people. The ability for sousveillant devices to be meaningful does not stem from the sheer complexity of their functions, nor the ingenuity of their creators. Instead, surveillance, as Foucault and Bentham discovered, derives its value from the power it has over the people who utilize it.

Chris Stross's 2007 novel *Halting State*, as well as the sequel *Rule 34*, are set in 2020s Scotland. A kind of wearable computing is depicted, which resembles the kind of contemporary cell phone usage commonplace in America today. Both books delve into the implications of a society which allows anyone to record anything at any time, especially in relationship to policing and authority in society. Lastly, the Posthuman Studios' role-playing game book *Eclipse Phase* showcases sousveillance as commonplace, because data storage technology and high definition digital cameras have become integrated into almost all objects. These three books may be seen as a cautionary tale of how quickly societies become consumed with sousveillance technologies before the rules or policies of usage have achieved consensus, or a shared understanding. As the last couple of years of Black Lives Matter has brought to the forefront, who has the right to record whom, and under what circumstances, are provocative questions in criminal justice, racial equity, and democratic representation. Sousveillance and alterity remain complex issues in the real world, not just in the pages of these futuristic books.

This genre of film also showcases sousveillance technologies. For example, in the 1995 science fiction thriller *Strange Days*, sousveillance recordings are made and sold as entertainment in the future. The film is set in a war-like Los Angeles in 1999. Foreshadowing contemporary struggles with racism, rape scandals, and celebrity voyeurism, the film's plot is centered around solving the murder of a celebrity by police officers. The murder is recorded by an observer who was secretly wearing an illegal sousveillant device This electronic device, called the SQUID, displays a sousveillant process wherein the recordings are made by a flat array of sensors that pick up signals from the brain stem within the cerebral cortex. The recorder headset, or sensors, are usually hidden under a wig, recording everything the wearer sees and hears. Recordings made while the person making them dies are called "blackjack" tapes and can be played back through a disc-like device, which allows the user to experience the memories and physical sensations of the original wearer or recorder. In the film, there is an active black market to sell SQUID recordings in the aftermath of people's deaths. In turn, the recordings function as something akin to how some watch reality lifestyle television in the present day. As this film reveals, however, the existence of sousveillant technologies can provide society much more than simple entertainment. If society chooses to cultivate technology to alleviate social problems, sousveillance may help us accomplish larger societal tasks, like building or destroying trust, solving crimes, and fostering alliances and networks within community.

One of the most interesting patterns in artistic representations of sousveillance is that of the pervasive invisibility of the various technologies. Both Foucault and Koskela (1979, 2004, respectively) assert that visibility is equivalent with power, meaning that those who watch are more powerful than those who are being watched. This shifts an often-accepted power dynamic that those who administer or enforce the rules may have greater power than those who are at the receiving end of such law and order. Additionally, these theories are important, because of the ways in which they manipulate visibility. It may seem valuable for resistance to be visible and overt to signal change and revolt, but this chapter offers examples of sousveillance in which the opposite seems to occur. Sousveillant technologies seem to conceal the resistance, or the technology is transparent or invisible, choosing to embed the technology in the skin, hide it under a wig, or conceal it under one's clothing as intimate apparel. Such invisibility may be conceived of as a kind of pragmatic democratic universality, encouraging all citizens to perform a civic task of monitoring, observing, and testifying to their everyday activities. Inconspicuousness of surveillance devices could be construed as the equivalent of the psychological internalization that Foucault cautioned against in *Discipline and Punish*. Here, Foucault argues that the awareness of being monitored by unseen observers can lead people to adopt the norms and rules of the surveillors. In turn, those being watched regulate their behaviours and ultimately begin to discipline themselves. At a minimum, sousveillance interrupts this process, and most severely, counter-technology proves Foucault wrong. For Foucault to be correct, instead of revolting, those being watched would have internalized the social norms being sought by the bureaucratic surveillance. What is intriguing, however, is that although resistance does occur, it is hidden or obscured. There is a kind of people power in most of these stories, because sousveillance technologies are often seen as great levelers of power given that most everyone in society has access to them. Yet, the power is not visible, as Foucault and Koskela argue it ought to be. This contradiction leads to an intriguing point for further analysis in civic society, as well as critical tests of the solvency of sousveillance. The place of sousveillant technology in civic societies demands citizens agree upon philosophical conditions concerning power. Indeed, sousveillant technology may prove appropriate, or effective, in certain societies, but not others. This variance stems from the diverse answers to the four foundational questions of sousveillant technology: (1) Must power be overt to be powerful? (2) What are the measurements and impacts of power in civic society? (3) Is it necessary, or even desirable, for the oppressed or surveilled to utilize the same tactics as their oppressors? (4) How do citizens hold government accountable or "check" abuses of power? Sousveillant technologies in artistic settings grapple intimately and creatively with these enduring questions, proving that American society can use a variety of sectors to help wade through civic debates. This section also provides a noteworthy reminder that US democracy should build alliances across sectors; that politicians and major corporations alone are not responsible for, nor even prudent in, determining the implications of power, security, and vulnerability.

CIVIC POLITY AND SOUSVEILLANCE

Sousveillance is characterized by emerging social networking communities which guard against, and interrogate, networked surveillance sites that are traditionally large, bureaucratic, and institutionalized. Facebook is a fascinating case study in this area, because while it is, in and of itself, a social networking community, it also has sustained severe allegations of fraud, spying, and stalking (Fernback 2013, p. 14). Hundreds of thousands of people using Facebook have joined online user groups geared at publicizing what participants believe to be Facebook's privacy abuses. Additionally, these groups seek to curb the spread of personal information in newsfeeds or from having personal data shared with law enforcement officials (Fernback 2013, p. 15). In a way, whether or not these advocacy groups are effective in achieving tangible change or solutions is moot. Arguably, the most pertinent part of their work, especially in a democratic society, is the push for transparency, the ability for advocates to openly critique the system or bureaucracy without fear of retribution, and the process of group deliberation and problem-solving. Facebook is one example of how people have been provoked by technology, but reflects a yearning for justice which goes deeper than that technology itself. Put differently, the real value of sousveillance may not be the technologies created, although they are powerful and fascinating. Instead, perhaps what is truly powerful is the people-power and the democratizing forces of resistance manifested in these technologies. Possibly the most revolutionary weapon resistance has ever had is the recognition of its own intelligence, empathy for others, and responsibility for improvement. Foucault (1979) claimed that "where there is power, there is resistance" (p. 95). An ever-constant partnership emerges in Foucault's vision in which power and resistance are ubiquitous. Therefore, sousveillance is best treated as a tool, like a microscope, in which power and resistance are put on slides and inserted for view to document, analyse, and extrapolate patterns. In other words, Facebook and other sousveillant technologies are not the answer, in and of themselves, but are helpful pieces within a broader discussion of how to construct a better civic infrastructure in America.

Several sousveillance advocacy groups that are interested in improving democracy through disclosure have sprung up with an interest in data-sharing. In response to calls for transparency, President Obama promoted three websites: (1) Government 2.0, which publicizes how public funds are being used and other issues related to fraud, abuse, and public policy; (2) recovery.org, which tracks spending and provides a geographic portrayal of where spending is occurring; and (3) data.gov, which provides access to all the databases used in public services (Peppers and Rogers 2009). Access to these websites is pertinent to legitimizing the activities of sousveillance advocates. It is not enough to simply record an incident perceived to be wrongdoing. After all, one must have access to the rules in order to determine if a violation or abuse has occurred. Thus, sousveillant technologies and informed citizenry are mutually dependent upon each other, at least at this early juncture.

One example of how sousveillance may foster socially compliant behavior—even though there is no obvious incentive—can be found in the field of medicine. Used in hospital settings, electronic monitoring can aid in audits and measuring compliance without the need for human supervision. In one medical study, for example, the term, Hawthorne effect, has emerged to describe a change in participant behavior (in this case handwashing to prevent hospital acquired infections) that is not attributable to incentives, or treatment regimens, but simply to the awareness of being watched (Adair 1984).

Given the ability of sousveillance to bring people together through the capture, storage, and transmission of personal experiences, society ought to more explicitly consider the potential of incorporating sousveillant technologies in democratic life. Effective communal life depends on understanding how sousveillance, and surveillance, at large, can help and hinder issues like dissent, resistance, anonymity, alienation, scapegoating, guilt, trust, and self-reliance. The proliferation of people-oriented technology like smart phones and YouTube is widely accepted as a defining feature of American life, and throughout much of the world, in the twenty-first century. Accordingly, the ubiquity of the technology molds all people into portable political theorists, positioned to consider whether this technological explosion is simply a widely successful fad of inventions, or if there are deeper, socially desirable or justifiable functions the nation's policies and values might connect with, and advance, by employing these technologies. These innovations may help citizens live more creatively and foster a stronger, more inclusive democracy. Moreover, and perhaps more fundamentally, the ways in which American society utilises sousveillance technology may reveal citizen relationships with power structures and what is desired in relationships with each other, as people and as citizens, in modes previously unavailable to the country in previous centuries. In the next section, this chapter will further discuss potential risks in the use of sousveillance as well as opportunities for civic transformation.

RISKS AND OPPORTUNTIES IN SOUSVEILLANCE

The informal nature of sousveillance may be considered a threat, or an opportunity to society. In practical terms, sousveillance faces less impediments from traditional media's commercial distribution, restrictions and procedures. Production is much simpler with sousveillant technologies, diminishing the need for expert knowledge. In bringing the amateur nature of sousveillance to light, Mann notes that inappropriate use of sousveillance devices will be scrutinized, but this is because sousveillance errors are visible, unlike those in surveillance cultures which are more secretive and unlikely to allow errors to be publicly detected. Here, Mann does not intend to deem which system, sousveillance or surveillance, quantitatively is in possession of more inappropriate activity. Instead, he means to suggest that the inherent foundational structure of sousveillance—with the focus on peer relations, personal agency, and trust—enables the public at-large the ability to find errors more readily. Other researchers have responded to privacy breaches arising from issues of surveillance as empowering. One researcher, for example, found webcams and exhibitionist acts enabled by technology to be a vehicle that can transform surveillance regimes into a kind of "spectacle" which emboldens citizens (Koskela 2004, p. 208). While empowerment might be a surprising result, it is an important one to consider in the realm of civic awareness.

In an example described by Bakir (2010) of the incidents from the 2003 Abu Ghraib prison scandal, American society was taken aback by the perceived inappropriate use of sousveillant devices. Although Baker provides systematic evidence that reveals a long history of abuse, which I will not go into during this chapter, the actions portrayed in American popular culture and mainstream news are germane to this chapter. Susan Sontag

(2004) captures the moral repugnancy necessary to understand the gravity of the situation:

> If there is something comparable to what these pictures show it would be some of the photographs of black victims of lynching taken between the 1880s and 1930s, which show small-town Americans grinning beneath the naked mutilated body of a black man or woman hanging from a tree. The lynching photographs were souvenirs of a collective action whose participants felt perfectly justified in what they had done. So are the pictures from Abu Ghraib. If there is a difference, it is a difference created by the increasing ubiquity of photographic actions. The lynching pictures were in the nature of photographs as trophies — taken by a photographer, in order to be collected, stored in albums; turned into postcards; displayed (paragraphs 6 and 7).

Photos of Abu Ghraib prisoner abuse were not taken by photojournalists or news reporters. They were taken by personal digital cameras, like a Cyber Shot camera, of U.S. soldiers perpetrating the abuse (Bakir 2010, p. 88). In a sousveillance culture, individuals determine what they find atrocious or beautiful. Individuals determine what is appropriate to record, which images to share, and to whom the information should be shared. As the previously discussed scandal demonstrates, sousveillance culture is not simply about the innocent fun of filming a child's first step and emailing it to grandma, or posting an inane quiz result on your personal Facebook wall to schmooze with virtual friends. Sousveillance culture bears considerable implications for political and legal conditions in democratic societies. While the Abu Ghraib example may be excluded from analysis for some, or given special conditions, because it is arguably rooted in wartime mentality, the examples throughout this chapter—whether in prisons, hospitals, the arts, and organizations or workplaces—demonstrate the pervasiveness of sousveillance. This raises important questions related to trust and consent.

The appeal of using sousveillance is reflected by the theory of discursive amnesia. This theory examines the public, or collective, forgetting, downplaying, or decontextualizing of economic or political events, which call into question a country's national actions for the sake of individual well-being. As Lee and Wander (1998) describe, in discursive amnesia "specific acts of collective forgetting that perpetuate privilege and interest in a particular economic and political context"; through these, "a group identifies itself, not only through what it publicly or officially recalls, but also through what it systematically forgets" (pp. 152, 154). Sousveillance

can appeal to activists and others who seek to remember, or raise, awareness of various issues in America, in turn amplifying voices of dissent in civic
life. Yet, just as there are challenges with "motivated forgetting" (Lee and
Wander 1998, p. 153), or deliberate squandering of political activities,
there are important concerns intrinsic to sousveillant technologies, as well.

Katina Michael (2015) explores the issue of consent at length, providing
a thought provoking challenge for sousveillance. If technology, namely
wearable cameras, record everything, such devices will inevitably capture
people in the recording process who have not granted their consent to be
recorded. Michael considers this issue to be at the level of a human rights
violation, suggesting that an individual has the right to control their own
image and go about their life without it being recorded. It is unclear from
her article if Michael holds surveillance societies to the same standard, but
it would seem both sousveillance and surveillance are at risk for violation of
the same infringement. This provokes several critical considerations,
beyond the initial fundamental, yet challenging, discernment of whether
Michael is correct in her baseline assumption. Her critique also raises legal
questions, such as whether governmental or authority figures, or even
advocates functioning within extreme oppression of their civil rights, are
justified in violating human rights for the sake of the greater good.
Additionally, Michael's critique suggests a prima-facie respect for the
individual's rights, but given that the individual is living among many other
individuals within a communal society, might collective rights play a more
substantial role? Lastly, Michael's critique unveils an interest in human
freedom in the face of technological and machine power. The status quo is
engulfed with technology, arguably on a level never experienced before by
humanity. Humans' relationship with these forms of power is just as
complex and uncertain as our relationships with each other. Such perennial
issues lack easy, homogeneous solutions, but as democracy evolves in the
United States, the intrinsic connection between sousveillance and civic
society clamors to be addressed.

Sousveillance provides the technological ability for more people to
engage in strategic political communication. In a sousveillance society, the
state no longer claims a monopoly on political communication, nor are
exposes relegated to a small number of media elite. Critically, however,
sousveillance is primarily focused on the technology. The legal system and
public sphere may be influenced by technological developments, but are
not determined by them. Sousveillance puts the mechanisms in place to
transform popular societal demands and moral outrage into real-world

changes or policy solutions. Yet, at the same time, American society must pay due diligence to the pressing moral and democratic questions which frame the culture in which sousveillance technology lives. Many issues loom over this chapter regarding resistance and social change. While it is not the intention of this chapter to address these enduring philosophical issues, it should still be noted that the fate of surveillance is inextricably tied to the democratic values and civic virtues held by American society. The technological power afforded by sousveillance will yield no social change, or worse, will harm society, if it is allowed to proliferate in a society with unclear moral codes or social expectations. While sousveillance may be able to hold power-abusers accountable, without this secondary focus, the full potential of it is unlikely to be realized. Referring back to the Abu Ghraib prison scandal, some analysts have put the best construction on it by suggesting that the incidents provoked "prolonged political introspection within the US on the practice and morality of torture" (Bakir 2010, p. 162). The grotesque trauma sustained in this scenario is only grotesque if viewed through the lens of social responsibility. It is not enough for us to have technology without moral analysis. Likewise, it is not responsible to allow technology to overrun our moral analysis.

Implicit in the preceding cautions is an even older challenge proclaimed in the pragmatic philosophy of John Dewey. Commenting in the aftermath of World War Two, Dewey stated in 1950 that "living as we now do in what is almost a chronic state of crises, there is danger that fear and the sense of insecurity become the predominant motivation of our activities" (Dewey 1993, p. 247). He explicitly connected fear-ridden choices with civic polity by suggesting that "when we allow ourselves to be fear-ridden and permit it to dictate how we act, it is because we have lost faith in our fellowmen—and that is the unforgiveable sin against the spirit of democracy" (Dewey 1993, p. 248). Although Dewey was writing about a society far less surveilled than present day America, the trends of commercialism, patriotism, and militarism, in which his critique is situated, remain pertinent today. Dewey's critique underscores the need for fervent policy reforms that respond to the deep-rooted causes of our country's troubles. In the spirit of Dewey's humanism, sousveillance must not stem from fear or insecurity, because this will not fulfill any kind of democratic promise. While recording people whom one is suspicious, or distasteful of, may be tempting, such motivations could make sousveillant technology appear as simply one more method to instill divisiveness and mistrust in American society. This chapter echoes Dewey's vital call for democracy to fulfill its promise and underscores the necessity of considering the motivation behind new technologies and policies. The desire for individual expression and protection must be balanced with the collective will. Sousveillance has the potential to provide new tools in America's progress toward social justice, but it also, in its current infancy, could be sidestepped, diluted, or impaired if not carefully explored. The impact of invention on Americans is felt across communities. As such, communities of all kinds must engage proactively and with compassion to consider how our interactions are structured.

The scale of conversations which stem from sousveillance are wide and important. Since the first technological inventions of humankind, people have discussed how technology can serve as a gateway to larger sociopolitical and ethical conversations about how to live fuller, more efficient, and more compassionate lives. In a word, technology is seen as an avenue to making life "better". Yet, perhaps more significantly, sousveillance reflects a struggle for culture, provocatively examined by Marvin Harris in his book *Cultural Materialism* 1979. As this chapter describes, the characteristics and challenges embedded in the creation and application of sousveillant technologies are consistent with Hariss's description of the American Museum of Natural History curator Clark Wissler's early twentieth-century definition of the three divisions of culture: "material traits, social activities, and ideas" (Harris 1979, p. 279). Undoubtedly, sousveillance readily fits into a discussion of cultural creation, but it is just a piece of the conversation. Put differently, sousveillance would be unlikely, in and of itself, to be capable of reshaping the dynamics of power in American society. All the same, sousveillance provides tools which have been unprecedented in human society and present enormous opportunity to organise, promote, and broadcast thoughts, opinions, behavior, and stories. As a collective in society harnessing new civic agendas, people have an ability to creatively and compassionately reflect human nature, agency, and political struggles with a newfangled authenticity. Yet this opportunity may not be realized. Mann notes that there is nothing about sousveillance which prevents re-appropriation by the Panoptic, traditional surveillance state (Mann 2001). Moreover, although sousveillance was developed by Mann with the intention to resist the Panopticon, there is nothing intrinsic in most technologies that mandates, or controls, their usage exclusively for that purpose.

CONCLUSION

Sousveillance devices have already become a cherished form of technology for many Americans. Through these devices, millions of users enjoy entertainment and connect to their surroundings with newfangled ease to watch, record, and broadcast their routines. Indeed, sousveillance could be simply a fascinating exercise in personal identity and artistic exploration. Yet, this chapter has called for the full potential of sousveillance to be harnessed. People should use it to stand against unjust oversight of their activities, contravene abuses of power, and become more informed citizens due to the enhanced observational scope. Ultimately, sousveillance can hold greater significance in society as an innovative tool in civic engagement. As mainstream news sources reveal, sousveillance has already infiltrated American life through recordings of police brutality Sousveillance should be used as a tool in the nation's ongoing efforts to foster openness, creativity, and trust in civic life. Surveillance is ever present in the status quo, and sousveillance presents an opportunity to nurture the nation's cultural imaginary's ability to forge healthy relationships between institutional power, vibrant communities, and compassionate, aware, and engaged citizens. Much is still being explored in sousveillance, but the technology warrants a place in the nation's discussion of governance, on large and small scales.

REFERENCES

Adair, J. G. (1984). The Hawthorne effect: A reconsideration of the methodological effect. *Journal of Applied Psychology, 69*(2), 334–345.

Bakir, V. (2010). *Sousveillance, media and strategic political communication: Iraq, USA, UK*. New York: Continuum.

Bentham, J., & Bozovic, M. (1995). *The panopticon writings*. New York: Verso.

Brin, D. (1990). *Earth*. New York: Bantam Books.

Crowley, J. (1985). Snow. In J. C. Oates (Ed.), *American gothic tales*. New York: Plume.

Dewey, J. (1993). John Dewey responds. In D. Morris & I. Shapiro (Eds.), *John Dewey: The political writings* (pp. 246–248). Indianapolis, IN: Hackett Publishing.

Fernback, J. (2013). Sousveillance: Communities of resistance to the surveillance environment. *Telematics and Informatics, 30*, 11–21.

Foucault, M. (1979). *Discipline and punish: The birth of the prison*. New York: Vintage Books.

Harris, M. (1979). *Cultural materialism: The struggle for a science of culture.* New York: Random House.

Koskela, H. (2004). Webcams, TV shows, and mobile phones: Empowering exhibitionism. *Surveillance & Society, 2*(2/3), 199–215.

Lee, W., & Wander, P. (1998). On discursive amnesia: Reinventing the possibilities for democracy through discursive amnesty. In M. Salvador & P. M. Sias (Eds.), *The public voice in a democracy at risk* (pp. 151–172). Westport, CT: Praeger.

Mann, S., & Niedzviecki, H. (2001). *Cyborg: Digital destiny and human possibility in the age of the wearable computer.* Toronto: Doubleday Canada.

Mann, S. (2003). Cyborg logs and collective stream of (de)consciousness capture for producing attribution-free informatics content such as cyborlogs. *First Monday, 8*(2). Retrieved from http://firstmonday.org/ojs/index.php/fm/article/view/1030/951.

Mann, S. (2004). *Sousveillance: Inverse surveillance in multimedia imaging* (pp. 620–627).

Mann, S., Nolan, J., & Wellman, B. (2003). Sousveillance: Inventing and using wearable computing devices. *Surveillance & Society, 1*(3), 331–355.

Michael, K. (2015). Sousveillance: Implications for privacy, security, trust, and the law. *IEEE Consumer Electronics Magazine*, 92–94.

Peppers, D., & Rogers, M. (2009). *The Societal Benefits of Data-Sharing.* Retrieved from http://www.1to1media.com/View.aspx?DocId=31350.

Posthuman Studios, LLC. (2009). *Eclipse Phase Core Rulebook.* Retrieved from http://www.eclipsephase.com/releases/eclipse-phase-core-rulebook.

Sawyer, R. (2002–2003). *Neanderthal Parallax.* New York: Tor.

Sontag, S. (2004). *What have We Done?* Retrieved from http://www.24Mserendipity.li/iraqwar/susan_sontag_what_have_we_done.htm.

Stampler, L. (2014). This 'bra cam' shows how often women's breasts get ogled. *Time Magazine.* Retrieved from http://time.com/3449830/nestle-bra-cam-breasts/.

Sterling, B. (2006). Alberto Frigo. *Wired Magazine.*

Strange Days. (2004). Directed by K. Bigelow [Film]. USA: 20th century Fox.

Stross, C. (2007). *Halting state.* New York: Ace Books.

Stross, C. (2011). *Rule 34.* New York: Ace Books.

Wilson, M. (2015). For 11 years, this man has taken photos of everything his right hand touches. *Fast Company.*

Author Biography

Mary Ryan is a doctoral student in Social, Political, Ethical and Cultural Thought at Virginia Tech where she teaches in the Department of Political Science. Her writing has been published in the *Journal for the Study of Peace and Conflict* and twice in the *Nomadic Sojourns Journal*. She has a chapter in the upcoming *Representing the Other Half: Reflections of Poverty in Popular Culture* (forthcoming).

CHAPTER 13

Medical Surveillance and Bodily Privacy: Secret Selves and Graph Diaspora

Susan Flynn

Vision has held a privileged position as the sense most associated with notions of truth, knowledge, and power throughout the history of Western civilisation. The optical has laid claim to the "truths" of the surrounding world for many centuries. From the camera to the X-ray, technologies of vision have also been central to the sciences and the arts, articulating the watching and the knowing of others. These technologies of seeing, and of capturing what is seen, hold a peculiar fascination for humankind; what Sontag terms the "insatiability of the photographing eye" has enacted our hunger for knowledge and power and spilled into our own physical self-assessment. If seeing is knowledge and knowledge is power, we seek to see as a way to know and to own, as "to photograph is to appropriate the thing photographed" (Sontag 1978, p. 3). We now seek dominion over our own selves by intimately capturing and recording our health data. The camera, both in its traditional form and in all of its recent technological incarnations, has radically reconfigured relationships between viewers, images, and the world, creating indexical relationships and systems of understanding. Just as with the advent of cinema and television as new artistic practices and industries came the transformation of crowds into mass audiences; so too has the new biotechnical era precipitated a mass

S. Flynn (✉)
University of Arts, London, UK
e-mail: susan.flynn.1@ucdconnect.ie

© The Author(s) 2017
S. Flynn and A. Mackay (eds.), *Spaces of Surveillance*,
DOI 10.1007/978-3-319-49085-4_13

audience for health data. In this way, present technologies have expanded our fields of vision. Mc Luhan argued that new mediums extend our nervous system, and technologies can be seen as extensions of human faculties. Technological development has brought about not only a proliferation of images, but also a proliferation of different modes of vision, and with it numerous distinct relationships and dynamics that can be formed between the viewer and the image. There is, therefore, a complex relationship between viewer and the viewed, the observer and the observed.

Vision manipulates by "looking for", by reducing the infinite possibilities within the visual field to the order it seeks. Vision, therefore, can be considered as an act of domination which seeks to control the chaos of images. It seeks containment, boundaries, compartments, and taxonomies of phenomenal form, angles, homogeneity, stable relationships and above all, the security of binary distinctions (Hughes 1999, p. 168). The record of images, the photograph, scan or X-ray, is therefore tasked with articulating these taxonomies, becoming tasked with charting what is seen. Photographic technology has long been implicated in programs of surveillance and control, and contemporary society is unprecedentedly mediated by this exponentially evolving camera's eye.

SURVEILLANCE

Anxieties about surveillance and privacy are inextricably linked to this digital age. New technologies and surveillance capabilities continue to permeate a vast range of quotidian spaces. A number of theorists have noted the ways in which surveillance, once seemingly solid and fixed, has become much more flexible and mobile, seeping and spreading into many life areas where once it had only marginal sway (Bauman and Lyon 2013, p. 3). We are monitored aerially and internally; from satellites to smart watches. In addition, technologies such as pixel trackers and spyware are embedded in the virtualized (but not immaterial) spaces and practices of information and communication media. The Western focus on security is manifest in the salience of surveillance mechanisms where judgement is equated with rationality. As Lenton and Rose write: "Contemporary rationalities of security certainly seek to grid the spaces of existence with technologies for collecting and collating information, with algorithms for its analysis, criteria for judgement and strategies for coercive intervention (2009, p. 234).

Collated data forms statistical patterns which comprise the twenty-first century form of surveillance; we watch and are watched via the graphed data mined from our activities, health, social media interaction, online engagement and sousveillance details. We map our runs, speed, fertility, heart rate, calorie intake and diet, all the while sharing the intimate minutiae of our personal lives with the others in the graph. These others share our information, measure it against their own, and track their progress using us as a marker; the unseen other of whom we know so much. These surveillance users become our data diaspora; conscious of, and sharing our journey, using our surveillance as benchmark for their own success much as narratives of success abroad once reached the home shores. In the era of liquid modernity, mass surveillance and colonization of the private is alive and well (Bauman and Donskis 2013) and there is a rapid disappearance of privacy as it was once known. This violation of internal privacy is not a result of the totalitarian regimes envisaged in dystopian narratives, but an often democratic participation in mutual surveillance, rendered into data and transmitted via devices. From headsets that measure brainwaves to clothes that incorporate sensing devices, personal health monitoring has turned healthcare into a massive consumer market expanding far beyond the wearable sports and fitness monitoring devices which feature prominently in media.

Aside from the proliferation of healthcare apps, wearable wireless medical devices are seen to be reshaping healthcare and to offer the possibility of continuous physiological data in managing chronic diseases, or in monitoring patients' post-hospitalization. Google, in partnership with Novartis has patented and is about to begin trials on a digital contact lens capable of measuring blood glucose levels which could transform diabetic life (Reuters 2015). A growing number of other medical devices are becoming wearable, such as blood pressure monitors, ECG monitors and pulse oximeters. Technology, such as BodyTel, uses Bluetooth technology to allow patients to wirelessly send collected body data to their doctors, while Preventice Technology's BodyGuardian Remote Monitoring System and Avery Dennison's Metria Wearable Technology upload wearer data to the cloud. Textiles included in AiQ Smart Clothing bypass the need for add-on sensors by incorporating them in clothes to collect data. Bluetooth technology used in systems such as 9Solutions IPCS, tracks elderly patients' movements and sends data to caregivers. "Personalised and precision medicine" as it is known in the US and encouraged by Barack Obama's "precision medicine initiative" (2015) are part of a trend toward diagnosis via biological databases. This precision medicine market is predicted to be worth US87.79 billion by 2023 according to a report by Global Market Insights (healthcareitnews.com, 2016). In this climate, many pundits call for a radical revision of the ethics surrounding surveillance, knowledge and management of the body, referring to the power relations inherent in the clinical gaze. Is this clinical gaze a technology of control? Or, in a phenomenological approach, can we view patient data as a conscious and voluntary engagement with the social mediated world? In this view, surveillance of the body is an interaction rather than an occurrence on the body.

THE CLINICAL GAZE

The scopic regime of the nineteenth century which Foucault credited with the birth of the clinical gaze was based on optical examination through medical instruments; speculums, endoscopes and so on. The prestige of the clinical gaze was enhanced by the invention of a nosology and science of nosography: a system of disease description that made it appear that all illnesses fit within a definitive disease classification. In the classification of bodies through scopic technologies, the birth of the medical profession in the 18th century can, in a sense, be considered the first Big Data revolution; therefore, what we are witnessing now is the second Big Data revolution. Now, new medical surveillance technologies reconfigure the clinical gaze through digital monitors and through sousveillance technologies. The digital reconfiguration of the body as a space of management and control has altered, not just how the body is viewed, but also how the body is thought of. Indeed, as Bauer and Olsen (2009, p. 119) claim "the upsurge in electronic image production for clinical purposes has yielded an abundance of patient information that demands ethical consideration."

The impact of computing on contemporary medicine is immense, in fact leading to a disintegration of the traditional clinical gaze (ibid, p. 118). Foucault's notion of panopticism is now systematically employed at the state level and his notion of biopower is inculcated in the digitised data that is held about us on our phones, computers, bank cards, biometric passports, and in our medical records. The era of the all-seeing algorithm is upon us, as through sousveillance technologies, the proliferation of panopticism reaches into our most intimate personhood. The panopticon is now internalised by the data that the medical profession harvest about our bodies and by the data we harvest ourselves through sousveillance.

In contrast to surveillance, sousveillance is consciously employed and controlled by an individual for personal fulfilment and health and fitness purposes. A "quantified-self movement" now actively monitor and record their personal data, physical states and performance. Ubiquitous computing pervades personal and social life, tracing people through social-technical systems.

"Let's moor this in today's surveillance realities. More and more, bodies are, in an ugly but apt word, 'informatized'" (Bauman and Lyon 2013, p. 134), furthermore, we can see them as datafied. Recent reconfigurations of the medical gaze in biomedicine have brought surveillance into the body where the measurement of the self and its biological or physical activities is tracked and made legible. The surveillance of people now inculcates a notion of populations as a manageable and see-able mass; numerical profiling and data analyses now judge populations and the individual body as patterns are calculated, codified, rendered into probabilities and risks. The emphasis on the management of our own material selves; sculpting, optimizing, prolonging life and avoiding ill-health, fosters a fascination with health data. The breadth of cultural products; films, literature and art, some of which have been discussed in this collection, articulate this self-obsession and point to the salience of this medical self-gaze. Biomedicine reaches into our everyday activities,

> We understand ourselves in the language of biomedicine; judge ourselves in terms of the norms articulated by biomedical experts…look up our symptoms on the internet; check our disease susceptibilities with personal genomic tests and commercial body scans; think about reducing our risks with diet and exercise; worry—individually and collectively. (Rose 2012)

This form of self-understanding is a complex multilayer of beliefs, practices and feelings which interact with the fluidity of everyday subjectivity. Bauman and Lyon (2013) suggest that this is neither good nor bad; a form of adiaphorization which allows human choices even in the face of what may be considered to be ever decreasing ranges of choice. We consciously, for the most part, allow ourselves to be perceived by, and through, the discourse of data. The evolving surveillant assemblage captures voluntary flows of information about our bodies turning them into fluid and malleable "data doubles" (Haggerty and Ericson 2000). Data emanating from the human body is processed and calculated to create a data double so that data proxies for the actual human being. This voluntary surveillance, created by "logging in", enacts the human's choice while it simultaneously creates a shadow of the human; a graph double. The data double is trusted more than the actual human being and so software designers and professionals are deemed morally neutral by saying that they are "dealing with data" (Bauman and Lyon 2013). In the scientific reverence for rationality, the data is revered as more accurate, spared the anomalies and emotions of the human. This vision of human life has been enhanced by its convergence with big data, which is apparently unable to accommodate randomness. Disseminated and viewed in graphs, tables and percentages, these data sets comprise a new view of humanity and all of its potential risks. "Biology is now inextricably linked with information technology and has become in its own right a science of information. It is not interested in man, but rather in his elementary components" (LeBreton 2004, p. 2). Although Floridi (2016) argues that our data is inseparable from us, phenomenologically, we cannot be reduced to an algorithmic calculation or data set. Reverting to a sociological or phenomenological stance, the body is not reduced to a 'docile' surface when it actively engages in agentic action. The creative subject who utilises technology engages in interaction with technologies, allowing some of his elementary components (data) to be shared. The act of separating the human from his/her data may in this proffer a privatization of the individual, acting to preserve individual attributes in favour of calculated data.

BIOMETRICS

Biometric passports complicate the subject's agency in contemporary surveillance, as personal biological data is made a prerequisite of movement across borders. Biometry—the application of statistical analysis to biological data—harvests information from the body and plumbs it into surveillant information systems. Biometric passports, such as those used in the UK since 2006, feature a chip with the holder's facial biometric; a merging of information with biological data. A "real-time" management regime, biometric passports are the new "city gate", garnering insecurity as they propose to offer security. Lyon (2008) notes that the West prefers high-tech solutions, rather than low-tech ones, as they appear suitably dramatic and so are appropriate for dramatic risks (503). These are part of the system which attempts to be able to anticipate and tolerate disturbances and contingencies of security; a strategy for countering bio-insecurity in the context of national threat and risk management (Lentzos and Rose 2009).

After 9/11 the practices of security and risk management received much attention in the West; and terms such as "suspicious" became the new "bad". Insecurities appear just as fast as each new scanner at the airport, or at border crossings, and there is no knowing when we may be deemed a risk (Bauman and Lyon 2013, p. 101). Information held by state agencies has given rise to new open data debates.

In the health sphere, aggregates of patient data are used to monitor population health and form grids of potential health problems and patterns. State and organizational desire for positive identification merge in this field, where the measurement of bodies apparently collides with the prediction of characteristics in the normative project of crime prevention and population management. Biometric data is motivated by the objective to elude ill-health, fragility, frailty; the vagaries of the human body. A normative regime, it is a health risk technology which uses insurantial concepts based on the notion that risk is calculable according to probability. In biometric data, patterns can be identified, coded and made into probabilities. As hundreds of thousands of joggers upload their run data, a fitness graph is formulating in the cloud, surveillance of and by the joggers creating a data set to be read and analysed.

A new kind of citizenship is made possible through biometrics. Biological citizenship (Rose 2007, p. 132) refers to those citizenship projects that link their idea of citizens to beliefs about the biology of humans, as families and lineages, communities and species. Biological citizenship is comprised of data and statistics which are enmeshed with locales, hence constituting both exclusion and participation. Through biometrics, the body has become a locus for judgements and hence the locus for many of our discontents as well as a site of hope. Cultural expressions of this contradictory nature of bodies comprise dramas of surveillance, hope and redemption, ethical and unethical interference and optimization. As biomedical reach grows to assess what Rose calls "emergent forms of life", the surveillance of medical diagnostics becomes a regime of watching and judging:

> In and through such developments, human beings in contemporary Western culture are increasingly coming to understand themselves in somatic terms: corporeality has become one of the most important sites for ethical judgements and techniques. In this regime, each session of genetic counselling, each act of amniocentesis, each prescription of an antidepressant is predicated on the possibility, at least, of a judgement about the relative and comparative

quality of life of differently composed human beings and of different ways of being human. (Rose 2007, p. 254)

The age of biological citizenship in which we live is therefore replete with choices which we must make over our physical beings and those dependent on us. While the sight of, or knowledge of, various physical symptoms may be available to some of us, to the majority of the world's population, no such power is available. Furthermore, where biological data is concerned, objectivity is elusive. "The claim is often made that biometrics replaces the subjective eye of the inspector with the objective eye of the scanner, but the problem is that 'objectivity' is compromised by the ways that key differences—of class, race and gender—are defined" (Lyon 2008, p. 505). Biometrics calls into question the value of the real and lived experiences of the body; it is innately conservative and blind to the vagaries of experience and opinions as it reduces human life to the level of the numerical.

When questioning the cultural implications of using the body as a password, Lyon (ibid: 506) asks "what it means for identification to shift from stories to samples?" Surveillance is concerned with reducing persons to numbers; populations become data-sets, graphed and calculated along straight lines. Biology, in some senses a field of positivist knowledge, can be considered reductionist through the rise of the molecular style of thought which "analyses all living processes in body and brain in terms of the material properties of cellular components: DNA bases, ion channels, membrane potentials and the like" (Rose 2012, p. 2). The matter of the human body can now be seen as malleable at the most intimate level. Yet from the deterministic we have moved towards a probabilistic way of thinking about bodies and health (Rose 2012). The collected data forms a canon of probabilities from which we contemplate our options. In the milieu in which we enact our self-governance, data sets inform our lifestyle choices; we chose our actions with one eye of the graph.

Looking at data sets, codes and biometrics, do we see man? The virtual body, comprised of data, must fit into a pre-defined category. The "technologisation" of life means to know is to intervene; the greater capacity to act on biological knowledge means that life has become amenable to intervention and to projects of control (Rose 2012, p. 3). This "technologisation" of life is dependent on the medical gaze and the technologies of "looking". This surveillance regime is inculcated in the separation of human experience from the graph; data sets separate the material body from all of its lived experiences, thoughts and concerns.

The language of biomedicine is not subjective; levelling all of us in our clinical measurement. "These body identities permit classification and assessment based on 'samples' but exclude the possibility of hearing the voice of the person whose body is under scrutiny, in the form of 'stories' that she might tell … without them the 'body' is incomplete" (Lyon 2008, p. 507).

In this sense, biometric data is limited in scope and fails to provide a complete account of bodies and of real lives. In a reductionist sense, the increased salience of the molecular style of thought which analyses the human body in terms of material properties has resulted in a new language of bodies. The salience that the biomedical sphere has achieved in self-management is evident in the cultural language of surveillance. This body-watching forms what Rose terms the "somatic ethic" which is "disseminated through a network of injunctions from experts of the somatic; deemed to be a matter of state as well as the individual; and embedded in multiple sites from home and school to workplace and leisure"(Rose 2012, p. 4).

Predictive Analytics and the Currency of Fear

Strategies of control are now moving from reactive to predictive (Conrad 2009), seeking to use surveillance and sousveillance to prevent personal or social dangers. Effective prediction can be seen to be a positive outcome of such control systems, as big data predictions can lead to focussed preventative measures, such as Google's pandemic prediction in 2009. Google developed software which found a combination of forty five search terms, which when used together in a specific model had a strong correlation between prediction and real figures. This system proved a useful indicator which could predict and track in real-time the spread of disease which the Centre for Disease Control and Prevention could not do (Cukier and Mayer-Schonberger 2013). Heralded as a new era of disease control, this development merged big data analytics with contemporary cultural mores; the search terms and behaviors of a certain set of people in a certain place.

Predictive analytics are heralded as the new modus operandi of advertisers, social and medical experts. Everything is to be seen and known, for our own good. Patterns can be formed statistically, codified, made into probabilities and hence, suspicious anomalies identified. Predictive analytics are increasingly seen as having the potential to reinvigorate commerce and industry, and are being seized by all manner of social institutions. These institutions are thus able to predict whether we will "click, buy, lie or die" by predicting human behaviour (Siegel 2013). Such predictions, based on the big data mined from surveillance and sousveillance technologies, purports to combat risks and strengthen healthcare. Predictions are powered by the Western world's most potent product—data—which we scatter around us at every turn in the virtual and real world. Predictive models predict the behaviour of individuals by taking the characteristic of the individual as an input and providing a predictive score as an output (ibid). Characteristics are weighted; the attributes of an individual are factored together. The ethos of the technological world is thus employed in the grading and judgement of our individual characteristics, which are utterly decontextualized. In the medical realm, biometrics can be seen as an instrument of fear, the statistics becoming an essential component of our self-government. Biometrics can thus be seen to create fear; self-creation and self-assertion are tethered to comparisons with others, with the "standard" man.

> Fear, like water, has been made a consumer commodity and subjected to the logic and rules of the market. Fear has in addition been made a political commodity, a currency used in conducting the game of power. The volume and intensities of fear in human societies no longer reflect the objective gravity or imminence of menace: they are instead derivatives of the plenitude of market offers and the magnitude of commercial promotion (or propaganda). (Bauman and Donskis 2013, p. 102)

In the Western world, our maintenance as human beings has much to do with commercial agendas as "a global bioeconomy has taken shape around the manipulation of biology. Biological knowledge has become highly capitalised" (Rose 2012, p. 3). "The willing, nay enthusiastic, cooperation of the manipulated is the paramount resource deployed for the synopticons of consumer markets." (Bauman and Lyon 2013, pp.135–136).

Health surveillance systems are marketed to us via the imperatives of freedom to choose, of biological optimization, of security and of responsibility. We can observe these imperatives in cultural products such as many of the films and texts considered in this volume. Contemporary strategies for life, certainly in the West, incur self-regulatory measures, operating at what Foucault (2004) termed the "centripetal" level; within a circumscribed closed space (the body) one isolates and concentrates one's technologies (both material and biotechnological) and regulating all within that closed space (body). Conscious of the good and the bad within that space, one tries to correct each anomaly. Split from any consideration of morality, this is the ideology of our internal surveillance.

POSITIVE VIEWS OF THE MEDICAL GAZE

Our current regime of personal health surveillance has in these myriad ways been viewed by many of the mentioned theorists as suspect; separated from the greater issues of morality. However, new surveillance technologies do offer some positive developments which privilege material privacy. If the medical gaze is limited to data, then the individual material body is spared from becoming fodder for intrusion. The rapid development of imaging techniques and the new technological gaze now allow a greater material privacy to patients by avoiding disruption through visualization. Endoscopy, for example, looks through a tube inserted through the body while laparoscopy through a tiny incision; both intend to diagnose without major exploratory surgery. Digital imaging preserves the dominion of the material self; separating the biological body from the interior self. Bauer and Olsen (2009) point out that digitization of health care is married to non-invasive procedures; minimally invasive methods are facilitated by information technology and messy interventions are rendered obsolete thanks to digital imaging. A key benefit of this new scopic regime is that it allows much more focussed surgeries, thus resulting in increased bodily integrity. Key-hole surgery is made possible by such imaging technologies and bodily integrity is preserved, resulting in interference only at the exact position of the physical "anomaly". At the material level, therefore, such forms of surveillance proffer a protection of material human privacy and a parameter to the clinical gaze. Bodies may be protected against becoming areas of intrusion via physical examination and thus may be radically privatized from medical interrogation. As the body becomes digitally reconfigured, it is separated from its material form; the medical diagnostics centred on images and data rather than human flesh. The medical gaze is becoming de-localized in this manner.

Participation in sousveillance technologies can be seen to merge what Andrejevic (2005, p. 482) terms "the carrot of participation" with "the stick of generalized risk"; the need to enlist monitoring strategies as a means of taking responsibility for one's own well-being. "Thus, part of the promise of the interactive, information revolution is to provide the general public with access to the means of surveillance for do-it-yourself use in the privacy of one's home" (ibid). Although Andrejevic sees this as a neo-liberal form of governance in public and intimate realms, it proffers a democratization of knowledge and information, a retrieval of the medical gaze from professionals and an empowering ownership of one's own data. The "open data movement", gaining prominence in recent decades, seeks open access to data sets primarily held by state agencies, in tandem with the "right to information movement" (Kitchin 2014, p. 48). The open data movement seeks to radically transform ownership of and access to datasets, in the hope of democratising the production and distribution of "knowledge". The collection of one's own data offers self-knowledge, or a tool for the creation of knowledge. "Evidence-informed monitoring and decision-making" (ibid, p. 55) are heralded as the means to a fairer, more just world. The 'Data Box', one concept in development, aims to put peoples' data back into their own control. In collating one's own metadata, users would assert legibility and ownership of their own data (Haddadi et al. 2015).

In the community, home care of patients is facilitated by remote monitoring, and for a largely aging Western population, this may offer a realistic version of care of elderly or infirm patients in the future. Knowledge is shared among non-professionals as data about persons and "patients" is now diffused throughout society, through the grid of information networks, hence returning the power of knowledge back to society. Remote monitoring now offers a reconfiguration of patient-doctor relationships, moving medical care back into the realm of the patient's own environs. Patients may be relocated to homes from hospitals and care homes via the possibility of remote monitoring; a more sustainable and culturally sound manner of care and support among families and communities.

In Great Britain, DeepMind Technologies, an artificial intelligence company purchased by Google in 2014, has developed a healthcare app for medics. The Streams app monitors patients with kidney injury, tracking their data and providing medics with 'breaking news' style updates on their conditions. A 5 year agreement with three London hospitals to gauge the success of the app will begin in 2017. Data concerning approximately 1.6 million patients per year is expected to be shared with DeepMind, raising concern about the ethics of patient data use, although assurances have been given to the National Health Service regarding the sovereignty of patient data (Engadget.com). However, the time management potential and the real-time possibility for intervention may free up valuable care hours which could be allocated elsewhere.

In terms of global health, new monitoring technologies may identify global risks earlier and take appropriate actions. As the Google case (Cukier and Mayer-Schoenberg 2013) shows, health threats may be met with preventative measures which will protect humans in a more systematic and organised way and "new techniques for collecting and analysing huge bodies of data will help us make sense of our world in ways we are just starting to appreciate" (ibid, p. 7). The shift from the deterministic to the probabilistic may offer improved life chances via the prospect of prevention rather than intervention. Health data collection by individuals has the potential to result in earlier diagnoses and less invasive procedures as well as promoting healthier lifestyle choices.

Conclusion

First-world parents are now introduced to their new children through ante-natal 3D scans; the vision of new life is articulated through sound-wave surveillance of the foetus in the womb. At this very basic level of visualization, we see surveyed images of life as a vector of life itself. In this vein, surveillance is justified by sparing the material interference with the mother's body and the foetus as well as sharing the professional gaze with the patient through digital imaging; the audience is thus now privy to the clinical gaze. The popularity of 3D and 4D baby scans in the West points to the cultural reverence for surveillant technologies and the notion that images present definite information.

This form of surveillance, the omnipresent or "telepresent" ubiquitous monitoring with which the West is becoming so enamoured, reclaims the power of surveillance from power systems such as the clinical gaze. Sharing pictures of one's baby scan, for example, reclaims one's agency in this mediated world. Rejecting individual scrutiny by sharing and celebrating the minutiae of one's internal organs turns the look outward and redirects the clinical gaze from the "patient" outwards to society (Bauer and Olsen 2009). Culturally, we have grown accustomed to the image of, and the data about, a "person" and yet we know, too, that the notion of surveillant control can only reach so far into the interior of the person.

Bodies may be seen to be already "informatized" through communications and media systems, as algorithms track our real-time location, preferences, interests and purchases. The information produced can be extracted from and, seen as separate from the body that produced it as biometrics takes our bodily data and mines it into graphs and statistics, rendering some of us citizens and others suspicious. A surveillant technology in which we often readily participate, we share our data via sousveillant gadgets and applications, offering up our preferences, activities, performance as a measurable commodity. Yet in allowing us to offer up some information and keep some (our narratives) to ourselves, the subjective lives are protected. Such lateral surveillance, a form of peer-to-peer surveillance mediates the current "risk society", we remain part of society and its trends while donating only de-localized information. If we are, as some theorists suggest, part of new neo-liberal governmentality, we also maintain dominion over our innermost thoughts, stories and lives while we offer our data to the network.

References

Andrejevic, M. (2005). The work of watching one another: Lateral surveillance, risk, and governance. *Surveillance & Society, 2*(4), 479–497.

Bauer, S., & Olsen, J. E. (2009). Observing the others, watching over oneself: Themes of medical surveillance in society. *Surveillance & Society, 6*(2), 116–127.

Bauman, Z., & Donskis, L. (2013). *Moral blindness: The loss of sensitivity in liquid modernity*. Cambridge: Polity.

Bauman, Z., & Lyon, D. (2013). *Liquid surveillance*. Cambridge: Polity.

Conrad, K. (2009). Surveillance, gender and the virtual body in the information age. *Surveillance & Society, 6*(4), 380–387.

Cukier, K., & Mayer-Schonberger, V. (2013). *Big data: A revolution that will transform how we live, work and think*. New York: HMH Books.

Flynn, S. (2015). New poetics of the film body: Docility, molecular fundamentalism and twenty first century destiny. *American, British and Canadian Studies Journal, 24*(1), 5–23.

Foucault, M. (2007). *Security, territory, population*. London: Picador.

Haddidi, H., Howard, H., Chaudry, A., Crowcroft, J., Madhavapededy, A., & Mortier, R. (2015). Personal data: Thinking inside the box. *arXiv preprint* arXiv:1501.04737.

Haggerty, K., & Ericson, R. (2000). The surveillant assemblage. *British Journal of Sociology, 54*(1), 605–622.

Hughes, B. (1999). The constitution of impairment: Modernity and the aesthetic of oppression. *Disability & Society, 14*(2):155–172.

Kitchin, R. (2014). *The data revolution*. London: Sage.

Le Breton, D. (2004). Genetic fundamentalism or the cult of the gene. *Body & Society, 10*(4), 1–20.

Lentzos, F., & Rose, N. (2009). Governing insecurity: Contingency planning, protection, resilience. *Economy and Society, 38*(2), 230–254.

Lyon, D. (2008). Biometrics, identification and surveillance. *Bioethics, 22*(9), 499–508.

Mittelstadt, B. D., & Floridi, L. (2016). *The ethics of biomedical big data.* Oxford: Springer.

Rose, N. (2007). *The politics of life itself.* Princeton: Princeton University Press.

Rose, N. (2012). The human sciences in a biological age. *Institute for Culture and Society Occasional Paper Series, 3*(1), 1–23.

Siegel, E. (2013). *Predictive analytics: The power to predict who will click, buy, lie or die.* Hoboken: John Wiley & Sons.

Sontag, S. (1978). *On photography.* London: Allen Lane.

Web Sources

Retrieved December 2, 2016, from https://deepmind.com/applied/deepmind-health/streams/.

Retrieved December 2, 2016, from https://www.engadget.com/2016/11/22/nhs-deepmind-ai-app/.

Healthcare IT News. (2016). Retrieved August 30, 2016, from http://www.healthcareitnews.com/news/precision-medicine-market-skyrocket-past-87-billion.

Reuters. (2015). Retrieved August 30, 2016, from http://www.reuters.com/article/us-novartis-ceo-idUSKCN0R50E920150905.

Author Biography

Susan Flynn is a lecturer at the University of the Arts, London where she specialises in contemporary media culture, digital and body theory and media equality. Her work is featured in a number of international collections and journals such as *American, British and Canadian Studies Journal* and *Ethos: A Digital Review of the Arts, Humanities and Public Ethics.*

SPACES OF SURVEILLANCE: STATES AND SELVES
AFTERWORD

Professor Vian Bakir

There is a rising tide of 'veillance' awareness, *veillance* being Steve Mann's (2013) term for processes of mutual watching and monitoring by surveillant organisations and sousveillant individuals. In this final chapter, I sketch this rising tide of veillance awareness both in academic scholarship and society at large, noting that the experiential dimensions of veillance in specific times, places and contexts are rarely addressed. Yet, this question of *what it feels like to be surveilled and sousveilled* matters to diverse actors. It matters to global telecommunications companies seeking to ensure that their products and services do not offend social norms and remain privacy-compliant; to states seeking to balance the right to privacy with the need for security while exploiting the surveillance opportunities offered by global, digital communication flows; and to citizens, seeking to negotiate a complex veillance terrain that is decreasingly possible to disengage from and that affords multiple possibilities for sousveillance. Inevitably, this sur/sous/veillance nexus of mutual watching is reflected in artistic and popular cultural forms. Paying attention to these provides insights into our hopes and fears about possible futures offered by ubiquitous mutual watching, as well as how we make sense of, justify, and challenge past and present sur/sous/veillant practices. These insights are valuable both for the academic field and for all those involved in the social practice of mutual watching.

© The Editor(s) (if applicable) and The Author(s) 2017
S. Flynn and A. Mackay (eds.), *Spaces of Surveillance*,
DOI 10.1007/978-3-319-49085-4

A Rising Tide of Veillance Awareness

Academics from a broad range of disciplines have increasingly turned their attention to surveillance societies and surveillance practices. For instance, the Surveillance Studies Network and their academic journal *Surveillance and Society* (2002-) adopts a critical and sociological perspective frequently focusing on issues of social discrimination, social sorting and control by surveillance systems, as well as how people actually engage with surveillance. At the other end of the spectrum, coming from a more technical and practical perspective, the *International Journal of Monitoring and Surveillance Technologies Research* (2013-) shows how the disciplines of bioengineering, medicine, computer engineering, assistive technologies and smart surveillance systems benefit each other, especially when interpreting, analysing, and connecting processes, activities, mechanisms and components that contribute to the quality of life. Yet beyond seeking to make the surveillance society "better"—either through examining its social problems or seeking to make it function more usefully, fairly and efficiently —surprisingly little academic attention from the humanities has been devoted to questions of what it feels like to be surveilled:[1] hence the importance of this edited volume.

Also neglected by the surveillance literature are other processes of mutual watching, or what Mann (2013) terms "*veillane*". For Mann, there are multiple modes of veillance. *Surveillance* involves monitoring from a position of political or commercial power by those who are not a participant to the activity being watched or sensed: examples include CCTV cameras, sentiment analysis and programmatic tools used by marketing and state intelligence agencies. By contrast, *sousveillance* (Mann 2002) involves monitoring from a position of minimal power, and by those who are participating in the activity being watched or sensed. For Mann, sousveillance takes several forms. *Personal sousveillance* is a form of watching for self-knowledge and self-expression without political or legal intent (such as much social media usage, selfies, life-logging, quantified self and wearables) (Mann 2004). By contrast, *hierarchical sousveillance* has political or legal intent (such as when protesters use their mobile phone cameras to monitor police at demonstrations, or when whistle-blowers leak incriminating documents) (Mann 2004, Mann et al. 2003). Mann has also theorised efforts to avoid mutual watching. *Counter-veillance* is Mann's term for blocking both surveillance and sousveillance (Mann 2013, p. 7): examples include wearable technologies that detect and blind cameras, such as DJ

Chris Holmes' anti-paparazzi clothing that ruins flash photographs at night by reflecting light; or simply, total disengagement with networked technology by going "off-grid". *Univeillance* is Mann's term for blocking surveillance while enabling sousveillance (Mann 2013, p. 7): this is exemplified by default encryption recently adopted by big technology corporations like Apple and Whatsapp, thus encouraging people to communicate without fear of being surveilled by the state. Accepting the inevitability of surveillance in contemporary societies, Mann and Ferenbok (2013, p. 26) seek to counter-balance surveillance by increasing sousveillant oversight from below (what they term *undersight*) facilitated through civic and technology practices. Once this balance is achieved, they suggest that such a society would be *equiveillant*.

While we do not yet have a journal of Veillance Studies, we are starting to see a multi-disciplinary academic response to the contemporary condition of mutual watching. An important trajectory is Mann's work on veillance that started with his experimentation with sousveillant, wearable technologies in the 1980s and 1990s, and continues today at the University of Toronto's Electrical Engineering and Computer Science departments. Also of note is a forthcoming (2017) Special Issue on "Veillance and Transparency" in *Big Data and Society* (2014-), a journal that examines big data practices characterised by data qualities, such as high volume and granularity, and complex data analytics that link and mine disparate, unrelated data-sets for new insights: the Special Issue reflects critically on the consequences of the contemporary ubiquity of mutual watching in the digital realm post-Snowden for how societies are represented, realised, governed and challenged (Bakir et al., forthcoming 2017). This focus on veillance is a new, and I argue, useful departure from traditional surveillance scholarship as it allows us to better understand contemporary digital societies where the reach of state and commercial surveillance is global, interdependent and individually penetrative; and where people not only have little option but to participate, but many do so willingly, and even joyfully through their own sousveillance.

While not badged as veillance, other related scholarly discussions from the humanities include a literature on privacy that has much to say on issues of who watches whom in contemporary digital culture, paying particular attention to cultural practices, social norms, philosophical underpinnings and policy implications (McStay 2014, 2016; Fuchs 2015; Trottier and Fuchs 2014). Also relevant is the literature interrogating cultural and social practices of transparency (Bakir and McStay 2015; Woodbridge et al.

2012; Birchall 2011). Together, these literatures provide a rich milieu for scholars in the humanities to explore the experiential and identity-forming aspects of the contemporary social condition of mutual watching.

Outside the academic world, such questions about practices of mutual watching assumed renewed importance after revelations in May 2013 by national security leaker/whistle-blower Edward Snowden, a contractor for US signal intelligence agency, the National Security Agency (NSA) . Snowden told us of the reach and nature of contemporary state surveillance of citizens' digital communications in a wide range of liberal democracies, spear-headed by the NSA in partnership with signals intelligence agencies in the other so-called "Five Eyes" nations (Canada, Australia, New Zealand and the UK). While there have periodically been furorés about US state mass surveillance of American citizens in the 1970s (Church Committee 1976) and of EU firms from the 1980s–2001 (Piodi and Mombelli 2014; Schmid 2001), it was not until twelve years into the post-9/11 War on Terror that Snowden leaked an estimated 1.7 million intelligence files evidencing the extent and means of contemporary digital mass surveillance of citizens' communications by their intelligence agencies. His revelations focus on the USA and, to a lesser extent, the UK, but also include a wide range of other countries with intelligence sharing relationships with the USA. The leaks show that the data that intelligence agencies collect in bulk includes the *content* of communications (such as email and instant messages, the search term in a *Google* search, and full web browsing histories); and what is called *metadata* (in the USA) and *communications data* (in the UK) (for instance, who the Internet and telephony communications is from and to whom; when it was sent and duration of the contact; from where it was sent, and to where; the record of web domains visited; and mobile phone location data).

From 2013 onwards, a drip feed of published leaks revealed the richness and pervasiveness of the state surveillance. For instance, the NSA can record and store a foreign country's entire phone calls for 30 days thereby giving it a window into the past month (National Security Agency—Special Source Operations 2011a, 2011b, 2013a); the NSA and GCHQ (Government Communications Headquarters) spy on SWIFT and credit card transactions (National Security Agency—Network Analysis Center 2010); the NSA harvests hundreds of millions of contact lists from personal email and instant messaging accounts around the world, many belonging to Americans (National Security Agency—Special Source Operations 2012a, b, 2013b); the NSA collects almost 200 million text messages a day from across the globe along with the associated geolocation data, names gathered from electronic business cards, financial transactions made by linking credit cards to phone users, border crossings generated by roaming alerts and missed call alerts (BBC News 2014; National Security Agency 2011); the NSA harvests from intercepting communications huge numbers of people's images to use in sophisticated facial recognition programs (National Security Agency 2012); and the NSA spies on online porn habits to discredit Islamist 'radicalizers' through their sexual proclivities (National Security Agency 2013). Such NSA surveillance in bulk was justified as long as intelligence analysts cited a "foreign intelligence justification" (National Security Agency—Signals Intelligence Directorate 2011, p. 2).

The leaks also evidence the NSA's multiple covert hacking capabilities and intentions. For instance, it can covertly crack online encryption used to protect emails, banking and medical records (code-named BULLRUN) (Ball et al. 2013; National Security Agency—Cryptanalysis and Exploitation Services 2004, 2010). It can hack smartphones (Rosenbach et al. 2013; National Security Agency 2010a) with targeted tools against individual smartphones, such as being able to turn the microphone on to listen into conversations (code-named Nosey Smurf), high-precision geolocation (Tracker Smurf), stealthily activate a phone that is apparently turned off (Dreamy Smurf), and ensure that the spyware self-hides (Paranoid Smurf) (National Security Agency 2010b). It can globally implant malware on computers to enable their surveillance (National Security Agency—Tailored Access Operations 2013, 2008, 2006, National Security Agency—Technology Directorate 2011).

Since the leaks, intelligence agencies and their official oversight bodies have maintained that their mass surveillance programs are legal (with the exception of surveillance of US citizens' telephone metadata).[2] They also argue that mass surveillance is necessary—that a complete data set is required to enable discovery of new, unknown threats, as past information may help connect needed "identifiers" (such as telephone numbers or email addresses) and reveal new surveillance targets. This leads to a "collect everything" mentality (Intelligence and Security Committee 2015; Simcox 2015). Having attempted to collect it all, the intelligence agencies use a plethora of programmes to catalogue and analyse the data. Through big data analytics, disparate, massive data-sets are linked and exploited in real-time to derive insights and meaning—a practice common in commercial organisations to better target their wares to customers. While commercial entities avoid individual re-identification of people so as not to fall foul of privacy regulations, instead using big data analytics to deliver insights into groups of similar people rather than targeting individuals (McStay 2016), intelligence agencies see their task as prediction and pre-emption of security risks—especially the terrorist threat—and so individual identification is a key goal. For instance, the NSA relies on domestic and international metadata and "contact chaining" (National Security Agency —Signals Intelligence Directorate 2011: 1) to create large-scale graph analysis of some Americans' social connections, identifying their associates, locations at certain times, traveling companions and other personal infor-mation—described as "the digital equivalent of tailing a suspect" (Risen and Poitras 2013).

That such mass surveillance is facilitated by the major telecommunica-tions corporations through which so much of daily life flows is what makes this surveillance so pervasive and almost impossible to avoid unless people renounce all forms of digital communication. The leaks showed that communications content and communications data/metadata are collected in bulk from two sources. Firstly, the servers of US companies (via Planning Tool for Resource Integration, Synchronisation and Management (PRISM)) run since 2007 by the NSA. This is done in participation with the big Internet, computer, social media and telecommunications com-panies (Microsoft, Yahoo!, Google, Facebook, Paltalk, YouTube, AOL, Skype and Apple) although not necessarily with their consent, as they can be secretly compelled by the state to comply (Greenwald and MacAskill 2013). The second source of bulk collection is where the NSA and GCHQ unilaterally seize communications directly from the companies' private fibre-optic undersea cables, switches and/or routes carrying Internet traf-fic. As Snowden leaked this information, the big telecommunications corporations seemed genuinely unaware of this second practice. On its public revelation by Snowden, they expressed outrage at this significant breach of trust by the US and UK governments (MacAskill and Dominic 2013); Google, Yahoo!, Microsoft, AOL, Apple and Facebook sent the US Senate judiciary committee a letter calling for greater transparency and "substantial" reform of the NSA (Ball 2013); and Microsoft stated that it was consequently going to increase its encryption (Timberg et al. 2013).

Elsewhere I have considered the surveillance-sousveillance dialectic in light of the NSA surveillance revealed by Snowden, leading me to propose that the term that best describes contemporary (digital) societies where mutual watching has become the norm is *veillant panoptic assemblage* (Bakir 2015). Rather than allying myself solely to either of the two central metaphors used within surveillance studies—that of the panoptic and that of the assemblage—the *veillant panoptic assemblage* combines these metaphors while eliding them with the concept of veillance. The *veillant panoptic assemblage*, then, is a surveillant assemblage (comprising multiple commercial telecommunications companies as well as other companies through which data flows, such as finance and travel); data-fattened by our digital communication flows (including sousveillant forms ranging from selfies to whistle-blowing); appropriated by the panoptic state (intelligence agencies) for preemptive disciplinary purposes (such as identifying new targets of surveillance to help prevent terrorism); but that still offers the capacity for certain types of resistance to surveillance (for instance, the univeillance offered by encryption). I also suggest that the current iteration of the *veillant panoptic assemblage* is an arrangement of profoundly *unequal* mutual watching, where citizens' watching of self and others is, through corporate channels of data flow, fed back into state surveillance of citizens, with minimal reciprocal undersight of surveillant organisations and states by citizens.

What Does It Feel like to Be Surveilled, and Does It Matter?

So, how does the *veillant panoptic assemblage* make us feel? Following Snowden's revelations, political-intelligence elites regularly invoked public opinion as desiring greater security, and probably being prepared to give up privacy to enhance their security (Bakir et al. 2015). Fuelling this debate were a large number of public opinion polls conducted in the USA and UK. For instance, in the UK alone, in the two years following Snowden's leaks, 38 opinion polls on the subject of surveillance, intelligence services and personal privacy were conducted (Cable 2015). They largely find that what people find acceptable, or unacceptable, in terms of what surveillance powers the intelligence services have, and use, is fairly evenly split (YouGov 2014). They also show that people are more concerned about private companies sharing their data and tracking their behaviour than about government surveillance programmes (TNS 2014; Ipsos MORI 2014b); and that more (63%) trust the intelligence services to behave responsibly with their data (YouGov/Sunday Times 2015) than they do communications companies, Internet Service Providers, search engines and social media firms (54%) (Ipsos MORI 2014a). However, when the public was asked whether, if GCHQ had the resources to intercept/collect digital communications would the public trust them not to abuse this information, 52% believed that GCHQ could not be trusted whereas only 42% felt it could be trusted (YouGov 2015). While this plethora of opinion polls provides snapshots of people's trust levels in various corporate and state actors' surveillant powers, closed-ended opinion polls on complex and highly contextual issues of privacy, acceptability and trust are poor vehicles for addressing the experiential dimension of surveillance. Also, as privacy researchers observe, there is a disconnect between what people say concerns them and their behaviour, with high levels of concern expressed about online privacy, for instance, but little evidence of people doing anything tangible to protect their privacy online (McStay 2016).

Providing more insight than these opinion polls is an in-depth, participatory study, *Surveillance, Privacy and Security* (SurPRISE), of 2000 citizens from nine European countries (Austria, Denmark, Germany, Hungary, Italy, Norway, Spain, Switzerland and the UK) on attitudes towards surveillance-oriented security technologies and privacy (Pavone et al. 2015). This study involved large citizen summits conducted in 2014 (after Snowden's leaks) to generate quantitative data and to explore public views on these complex matters in some depth, allowing participants to hear a range of evidence before taking a measure of what they think. These studies focused on three security-oriented surveillance technologies: (a) Smart Closed Circuit Television, namely, digital cameras linked together in a system that has the potential to recognise people's faces, analyse their behaviour and detect objects; (b) deep packet inspection, which detects and shapes how messages travel on a network, opening and analysing messages as they travel, and identifying those that may pose particular risks; and (c) smartphone location tracking, which analyses location data from mobile phones to glean information about the location and movements of the phone user over time.

The SurPRISE study shows European public concerns about state surveillance; this is, perhaps, totally warranted given that NSA surveillance is primarily directed at foreign targets (that is, the rest of the non-American world) and given that there are extensive intelligence sharing relationships between the USA and key liberal democracies. Specifically, the SurPRISE study finds that the European public think some surveillance technologies are useful and effective for combating national security threats and should be used, but that acceptability varies according to whether the surveillance is of communications or bodies, and blanket or targeted. Surveillance of physical bodies (smart CCTV) and targeted surveillance of digital communications (smartphone location tracking) are more accepted than blanket surveillance of digital communications (deep packet inspection). Furthermore, the European public tends to reject security-oriented surveillance technologies where they are perceived to negatively impact non-conformist behaviour; and it demands enforced and increased accountability, liability and transparency of private and state surveillant entities alike (Pavone et al. 2015). The study also finds that, notwithstanding national differences, few people are willing to give up privacy in favour of more security (Pavone et al. 2015, p. 133). Rather, the European public wants transparent, accountable public agencies that inform citizens about their purposes and functions (Pavone et al. 2015). Yet, while these findings on the importance that citizens place on public accountability of state surveillance are valuable, again, this study fails to explore the experiential dimension of what it feels like to be surveilled.

Also of interest are the recommendations from an inter-disciplinary, multi-end user ESRC Seminar Series held across 2014-2016, *Debating and Assessing Transparency Arrangements: Privacy, Security, Sur/Sous/Veillance, Trust'* (DATA-PSST!). This Seminar Series examined how, post-Snowden, different aspects of transparency affect questions of privacy, security, sur/sous/veillance and trust. Across six full-day workshops DATA-PSST! enabled a wide range of actors interested in transparency to interact. These actors comprised Non-Governmental Organisations (NGOs), think tanks, defence consultants, technology firms with interests in privacy, politicians, national and international data regulators, national security whistle-blowers, news outlets, artists, designers and activists. Academics from many disciplines participated including: Journalism, Media, Cultural Studies, History, International Relations, Politics, Law, Philosophy, Sociology, Criminology, Religious Studies, Organisational Studies, Engineering and Computing. A range of topics were debated, some focusing on issues of mass surveillance, sousveillance and privacy, and others on issues of secrecy and accountability of political-intelligence elites. Clear recommendations emanating from these seminars include the need for greater digital literacy among the public so that they can make better sense of the scale and nature of the mutual watching that Snowden revealed. To enable this, the seminars also identified the need for greater engagement of artists and popular cultural forms to communicate, clarify and problematise contemporary practices of mutual watching (DATA-PSST 2015a, b).

With opinion polls and in-depth studies attempting to capture what people think of surveillance and privacy, but failing to explore what it feels like to be surveilled, and with the DATA-PSST! recommendations urging greater digital literacy and greater cultural engagement with contemporary sur/sous/veillance practices, this book is timely. Its inter-disciplinary humanities focus presents a number of ways of thinking about such questions deeply, critically and perhaps more imaginatively. Fundamentally, it offers insights into what it feels like to be surveilled in different contexts.

For instance, it poses the question: what does it feel like to be surveilled when one is a (witting or unwitting) participant in one's own surveillance? Lutton's chapter on portrayals of contemporary urban surveillance in Bret Easton Ellis's short story collection, *The Informers* (1994), observes the pervasiveness—but also the potential pleasure—of surveillant assemblages in the contemporary urban environment. I posit that these are questions that could usefully inform the shaping of smart cities as designers of urban architecture and their sensors and analytics programmes consider human desire, agency, temporal mental state, informed consent and ethics regarding what sort of data should be collected, where and when, and to what purposes it should be put. If someone walks through a public space, or into a commercial space, should they expect their presence and activities to be automatically surveilled? How should this vary if the space is one where privacy is normally assured (for instance, public toilets); or if the space is one that becomes potentially dangerous at certain times (isolated public toilets at night)? How should this vary depending on how the presence or activities of the surveillee is captured? Is evidence about surveillees' presence as registered by a CCTV camera more or less acceptable than evidence about their presence garnered by tapping into their wearable, biometric sensors? What if the surveillance captures, not just their presence and activities, but their intentions: for instance, is data about what they look at the property of someone else, if their looking happens to take place in a shop set up with sensors to find out what products their gaze lingers over? What about if their gaze lingers over people, and they happen to be life-loggers: do people want ubiquitous sousveillance, and if not, what norms and cultural codes will manifest to minimise the social friction that widely disseminated sousveillance technologies are likely to generate? (For greater discussion of such issues, see McStay (2016) and McStay (forthcoming).)

Most authors in this book find the answers to such questions to be quite dystopian. Pignagnoli's analysis of surveillance in post-postmodern

American fiction finds a dystopian society ruled by global surveillance, that claims transparency, that promotes a type of personal sousveillance (where people watch themselves and others in their lifeworld) but where this transparency is only of citizens/subjects/consumers to the surveillant state or corporation, and not one of reciprocal transparency of surveillant organisations to citizen scrutiny and undersight: in other words, personal rather than hierarchical sousveillance prevails. Pignagnoli suggests that these narratives are a reflection of the strong authorial/ethical voice in these fictionalised digital worlds. Certainly, some of these visions of the future resonate among those in the real world who also find current surveillant practices dystopian: indeed across 2014–2016, one of the subjects of Pignagnoli's critique, Eggar's novel, *The Circle* (referencing a fictional technology corporation that has bought out Facebook, Twitter and Google), was repeatedly cited in the DATA-PSST! seminar series by diverse 'end users' (national security whistle-blowers, privacy-oriented NGOs, and privacy-enhancing technology firms)—not for its literary prowess but for its seemingly close, if not already-realised, vision of a future without personal privacy or freedom.

The dystopian trope continues in Bacon's analysis of surveillance and the vampiric lens. Arguing that our response to ubiquitous surveillance is to replace one's real self by an inauthentic identity that is created for, and in, the observing lens/camera, he makes a comparison to the vampiric gaze as depicted in Bram Stoker's novel *Dracula*, Tod Browning's screen adaptation from 1931, and twenty-first century films. Here, the vampiric gaze acts as a form of soul training, as in Foucault's panopticon and its surveillant gaze: characters who do not comply with the social rules of the vampiric gaze fare badly, while those who conform to its social rules become transformed into the undead. While perhaps a far-fetched analogy with twenty-first century surveillance, it nonetheless provides food for thought about the gothic nature of surveillance that, unseen but felt everywhere, feeds off security and consumerism. Certainly, recent empirical research into the chilling effect of digital mass surveillance, where users change their search patterns online (Stoycheff 2016; Penney 2016; PEN America 2013), attests to the inauthenticity of the real self that is projected into the surveillant assemblage.

McBride's analysis of the contemporary techno-mediated modes of being in digital gaming culture (such as *Pokemon Go*) and wider literature and film also demonstrates a recurring dsytopian narrative—this time on the ever disappearing dividing line between human and machine, with attendant warnings about sinister hegemonic monitoring of everyday citizens and privacy loss. McBride explores the subtlety of the delivery of this dystopian narrative in Spike Jonze's mainstream 2013 hit film, *Her*, set in the near future; here, the type of visual focus adopted by the camera lens (predominantly shallow) reflects the main protagonist's shallow, ego-obsessed, narcissistic love for his digital personal assistant, who has shaped herself to satisfy his personality.

Other chapters suggest that the dystopian surveillance society is not dystopian enough, at least when it concerns American practices of racialised surveillance. Clapp's analysis of Claudia Rankine's poetry sees it as a reflection on the increasing proximity of surveillance society to the society of the spectacle, which work together in a new "aesthetics of transparency," one that hobbles classical liberal demands for political representation and personal expression. Also on the theme of racialised surveillance, *Tecle et al. examine* surveillance in three politicised spheres and social spaces in Canada across the 1970–1990s: higher education, hip hop music and radical activism. The authors focus on the psychic effects of surveillance on individuals and communities of Black people and advance the notion of "castration anxiety", namely the undeserved psychic suffering of those under the punitive eye of society. They argue that Canada's vaunted national integration policy of multiculturalism manages, excludes and psychically oppresses Black people. Their discussion of how Black empowerment was used to justify surveillance in the 1970s, and how the *Royal Canadian Mounted Police partnered with the* US Federal Bureau of Investigation to employ a black agent provocateur to monitor, discredit, and ultimately destroy Canada's Black Power Movement via surveillance has particular resonance given Snowden-revealed mass surveillance for targeting broadly defined threats to national security. It is a timely reminder that, historically, the security services of liberal democracies have used surveillance for socially problematic ends.

Moving beyond warnings of possible dystopian futures or near-futures offered by popular cultural texts, Pheasant-Kelly's analysis of surveillance in the film *Zero Dark Thirty* (Bigelow 2012@@) examines the realism of this filmic depiction of real world events regarding the hunt for Osama bin Laden. She considers how these reflect changes in real-world monitoring of both public and terrorist activities since 9/11, in particular, its legitimisation of the intrusion into privacy through visual and digital profiling surveillance technologies. Of course, the film's realism and legitimisation of surveillance (as well as of Enhanced Interrogation Techniques) can be explained by the fact that the Central Intelligence Agency (CIA) was involved in developing and vetting the film's script—this revealed by Freedom of Information Act requests by *VICE* news, (Leopold and Henderson 2015). Interestingly, out of all the cultural artifacts examined in this edited collection, the one directly influenced by the CIA is the only positive depiction of surveillance.

Moving beyond purely dystopian analysis, other contributions to this volume observe the utopian as well as dystopian potential of sousveillance. Flynn's analysis of medical surveillance and bodily privacy suggests that new medical surveillance technologies reconfigure the clinical gaze both through digital monitors and through sousveillance technologies; she suggests that via sousveillant health and fitness gadgets and apps, our offering up of our preferences, activities and performance as a measurable commodity (in the form of data and big data) spares the individual material body from becoming analysed, thereby protecting our subjective lives. Yet, while this is certainly the sales pitch from those seeking to leverage health and financial value from big data, there are ongoing debates within the wider fields of big data and privacy, especially regarding anonymisation, pseudo-anonymisation, de-identification and re-identification of data (for instance see McStay 2016). These debates suggest that we would be wise not to assume permanent benevolence from big data corporations given the nature of healthcare around the world; while in the UK, health is typically seen as a public good, elsewhere health is clearly a business, and thereafter linked with the insurance industry, financial risk reduction and the desire for transparency of the subject. The risk of individual re-identification, then, remains a real and long-standing issue.

Ryan's analysis of sousveillance as a tool in US civic polity explores the potential of specific sousveillant technologies to improve American democracy by producing engaged citizens through their recognition and enactment of people-power and the democratising forces of resistance. For Ryan, sousveillant art, such as that produced by life-logger Alberto Frigo, both document and explain people's individual experiences; while that produced by Steve Mann function as protest art and social commentary. Both place an individual in control of how they will be seen, if at all. Ryan also explores the representation of sousveillant technologies like wearable computing in science fiction books and film, these providing both cautionary tales of how quickly societies become consumed with sousveillance technologies before established norms of usage have been achieved, but also how such technologies can help alleviate social problems like solving

crimes. While Ryan suggests that a prime benefit of sousveillant technologies is to encourage all citizens to perform a civic task of monitoring,
observing and testifying to their everyday activities, she acknowledges the
problem of panoptic reappropriation: the *veillant panoptic assemblage*, as
discussed earlier in this chapter, then, remains a concern. Elaborating on
the theme of panoptic reappropriation, but expanding on how sousveillance may be conceptualised, Milligan's analysis of DeLillo's novel, *White
Noise*, highlights how we currently attempt sousveillant connections
through our mobile and gaming technologies to the corporate surveillance
around us. He explores our sousveillant desire to connect with those gazes,
for instance through gaming avatars where we create our ideal self (a form
of personal sousveillance), which we then invite or allow others to surveil:
we would rather be watchable than just be watched.

While post-Snowden, the response of technology corporations has been
encryption to resist covert state surveillance of digital communications,
several chapters in this book suggest alternative artistic responses. Artists
have variously sought to enable individual empowerment and agency in
self-representation by repurposing surveillant systems and tools; and by
using webcams as performative prosthetics to the human body to complicate the entangled representations of real-time and real-place. Christmas'
chapter questions how the dystopian metanarrative of the surveillance state
and passively docile subjects, apparent in art, novels and academic critique,
can be challenged. She focuses on American multimedia artist, Jill Magid,
whose work hijacks and repurposes techniques normally allied with
surveillant technologies (such as CCTV cameras). In allowing the subject
to reclaim agency in their experience of surveillance, Magid empowers the
subject in their self-representation. Meloche's chapter on digital performance art using webcam prosthetics highlights the real-time experience of
the artist/subject, rather than fine art's focus on objecthood. I would
further pose the question (following Rancière 2011 [2008]): if artists can
subvert the surveillant gaze, can ordinary members of the public, or is the
legacy of artistic intervention simply to raise critical questions in the minds
of the public? Perhaps simply raising critical awareness is enough, given the
rarity of in-depth citizen summits asking people to think deeply on issues of
veillance.

To conclude, this collection primarily shows that popular culture
responds to ubiquitous surveillance in a dystopian register. It is not a
condition that seemingly sits well with us, despite the potential of
surveillance to preempt or solve social problems like crime or terrorism,

and to make cities and health services function better; and despite the fact that certain forms of watching (such as sousveillance) have been embraced by the masses who use social media as well as the more niche life-loggers and quantified selfers. Yet, given the rise of sousveillant technologies, and the centrality of easily-surveillable digital communications through which we live so much of our lives, it is also a condition that we cannot easily escape, while its often abstract nature enables the practice to fade from consciousness. Artists, both of sousveillance and surveillance, attempt to problematise surveillance and show how to regain control of our own personal epistemologies. Cumulatively, how these artistic and popular cultural forms address contemporary practices and possibilities of veillance contributes both to public literacy and collective imaginaries on such issues. Citizens, businesses and policy-makers alike could do worse than to explore these popular and cultural artifacts while reflecting on what policy, product or personal changes could avoid the dystopia. Veillance studies, as an academic discipline, would also do well to more directly engage with the experiential aspects of mutual watching, so as to better understand what aspects of the sur/sous/veillant society are of most concern and of most delight to people, and to better advise on how increasingly integrated systems of mutual veillance can be designed to ensure that the dystopia does not prevail.

NOTES

1. This question has received considerably more attention from the social sciences: for instance, see Manolo et al. (2016), McEwan (2013), Ball (2010), Essén (2008).
2. Accordingly, on 2 June 2015, the USA Freedom Act was passed, restricting bulk collection of telephone metadata of American citizens, although not of foreigners.

INDEX

© The Editor(s) (if applicable) and The Author(s) 2017
S. Flynn and A. Mackay (eds.), *Spaces of Surveillance*,
DOI 10.1007/978-3-319-49085-4